AGRICULTURAL PROBLEMS IN ETHIOPIA

AGRICULTURAL PROBLEMS IN ETHIOPIA

HARJINDER SINGH
Department of African Studies
University of Delhi
Delhi

Gian Publishing House
29/6 SHAKTI NAGAR • DELHI-110007 • INDIA

GIAN PUBLISHING HOUSE
29/6 Shakti Nagar, Delhi-110007

ISBN 81—212—0078—4

PRINTED IN INDIA
Published by Mrs Gayatri Garg for Gian Publishing House, Delhi (India).

Printed at Mohan Printing Corporation (Composing at Godwall Printers), Delhi.

Preface

ETHIOPIA has historically been self-sufficient in food. During the Second World War, the country was reported to be potentially the 'bread basket of the Middle East' and it was in fact net grain exporter. Despite a long history of self-sufficiency, the country experienced crop failures repeatedly and consequently hardship to the population. Famine of 1972-73 led to the death of hundreds of thousands of Ethiopians. It has been estimated that about one-fifth of the population of Wollo province had died from starvation and the total number of death from the starvation might have reached 500,000 in the country.

Ethiopia was of course the first African country to introduce economic planning. The First Five Year Development Plan was launched in 1957. But the plans did not specify in detail the targets and achievements of various sectors. Regional projects were inadequately planned. The completion dates of all the plans were extended time and again. Funds for various programmes were allocated in a generalised fashion. Most of the funds were spent on unproductive items such as the army, the police and the courts. For example, the budget of 1973-74 shows that about 39 per cent of the budget was for these three items and the expenditure on productive sectors such as agriculture, industry and commerce was only 3.8 per cent.

It is said that the poor performance of agriculture during the pre-Revolutionary period i.e., before 1974 was mainly due to drought which affected agricultural sector adversely. With drought the area under irrigation had been reduced, available pasture land for livestock devastated and the sources of water for daily use dried. But the loss of so many lives in Ethiopia was not simply due to lack of rainfall. Other factors besides the drought too have contributed to the lethal proportions of the famine.

Ever since the inception of the Revolution of 1974, the first task that demanded most of the time of the Military Government was the endeavour to accelerate the growth of agriculture. It nationalised all rural land and distributed it to the tillers upto 10 hectares. Co-operatives have been established to lay the basis for socialism. Such reforms have been regarded as quite radical as they have brought land distribution and organisational changes. Despite these drastic changes, the country once again underwent a prolonged period of drought and famine in 1984 with a severity that approached that of 1972-73. It affected about one-fourth of the population. With the continuous fall in agricultural output in recent years it is likely that another human tragedy of vast proportion may occur. If these falling trends are not reversed, the future of Ethiopia is in jeopardy.

The task set in this book is to assess the performance and strategies followed for the development of the agricultural sector during the pre and post Revolutionary periods and pinpoint various constraints to its development. Part I starts off with a brief discussion of the place of agriculture in the Ethiopian economy followed by socio-economic characteristics of the country. Part II deals with the analysis of trends in agricultural output, area and yield during the pre-Revolutionary period and highlights the basic problems of land tenure, agricultural credit and marketing organisations. The community development programme which was introduced to improve agricultural situation in the country also could not make much contribution. The reasons for the failure of this approach have been discussed in this part.

Part III deals with the changes in the agrarian structure after the Proclamation of Agrarian Reform of 1975 and its impact on agricultural output during the post-Revolutionary period. This part also highlights the problems faced by the government, various organisations and peasants in the implementation of land reform programme.

A study of this kind requires an extensive original research into the primary source of information. This is particularly important in view of the paucity of written material on agricultural aspect. I visited Ethiopia in 1969 to study agricultural situation during the three plan periods, i.e. 1957 to 1973. The trip was financed by the University of Delhi. Since most of the material on this aspect was not available at one place, I had to visit the offices of the Ministries of Agriculture, Land Reform and Administration, National Community Development and Social Welfare and Planning Commission. I also took the opportunity to discuss agricultural problems with the Director, Institute of Ethiopian Studies who offered valuable and constructive suggestions.

I also visited some of the places near Addis Ababa to observe agricultural practices being followed by the farmers. I went to some of the main markets to study the marketing system. These visits helped me in understanding the basic problems of agriculture during the pre-Revolutionary period. The analysis of agrarian reform and its impact on agricultural output as well as on the economic conditions of peasants is mainly based on the government publications and the various studies conducted by individuals and institutions since the Revolution.

Finally, I would like to mention my gratitude to those who helped me in the production of this book. I am specially indebted to my teacher and now colleague Professor Daleep Singh who gave valuable guidance and suggestions. I offer my thanks to my other colleagues, Professor V.P. Luthera, Mr. K.B. Bajpai, Head of the Department, Mr. K.K. Virmani, Mr. O.P. Goel and Mr. R.P. Sood, Documentation Officer with whom I discussed some important aspects of the problems examined here. I am also deeply indebted to my friend Mr. Attar Chand, a well-known writer, who encouraged me to write this book.

I am particularly grateful to my wife Surinder Bir and children for their patience and constant encouragement during the period of writing this book.

Department of African Studies,
University of Delhi,
Delhi-110 007.

HARJINDER SINGH

Contents

PART I

THE SETTING

CHAPTER ONE

Introduction

THE DEVELOPMENT problem of the less advanced countries of the world is basically that of converting a 'traditional' society into a 'modern' one. A traditional society is predominantly based on subsistence or near-subsistence agriculture in which the per capita income grows slowly or may even decline as a result of population pressure. It has been estimated that nearly half of the world's population falls in this category.[1] In a modern society the growth of per capita income is internalised in the social and economic system through an automatic mechanism promoting capital formation and improvement of technology. In more specifically economic terms, the establishment of a modern society requires 'Industrialisation'.

Following this line of thinking some writers, such as Albert Hirschman agreeing with Karl Marx, have suggested establishment of industries without stressing the need for revolution in the field of agriculture at the early stages of development.[2]

1. John W. Mellor, The Economics of Agricultural Development, Cornell University Press, New York, 1966, p. 3.
2. Albert Hirschman, The Strategy of Economic Development, New Haven, Yale University Press, 1964, and M.M. Bobbler, Karl Marx's Interpretation of History, Cambridge University Press, 1950.

They insist that a developing country has limited resources for investment and that it cannot do everything at once. It has therefore to invest in that industrial sector which gives immediate and profitable returns. The theory that a developing country should develop its agricultural sector and export primary products in exchange for industrial products, they say, is a spurious one, for there is a long-run tendency for prices of primary products to fall in relation to prices of industrial and manufactured goods. Therefore, the prices a developing country would have to pay for industrial products would constantly rise, while the prices of agricultural products would comparatively be low, thus placing the underdeveloped country at a disadvantage. Development of the industrial sector would produce skilled, energetic, progressive-minded people by whom "a spark will be kindled which will ignite the rest of the population."[3]

This type of approach to economic development was adopted by the Soviet Union, Communist China, Poland, Argentina and Yugoslavia. When we look at the experiences of these countries towards their economic development, we find that all of them were confronted with difficulties midway during the implementation of their plans. Communist China inaugurated a planned approach to economic development in 1953 giving a high priority to the location of heavy industries at the initial stages. But at the time of its most severe food shortage it had to admit, and still maintains today, that agriculture was the basis of its economy.[4] The result was that in the middle of the First Plan some of the industries were closed following non-availability of inputs from agriculture. The Chinese Government modified the Plan and shifted the emphasis to agriculture giving it higher priority.[5] Similarly, Argentina too, did not succeed in her attempt to industrialise the economy

3. *Ibid.*, p., 180.
4. United Nations, African Agricultural Development, New York, 1966, p. 17.
5. *Ibid.*, p. 17

speedily and found it necessary to import wheat from the United States. Yugoslavia and Poland also faced the same difficulties due to the lack of necessary inputs from the agricultural sector.

It is true that the deliberate sacrifice of agriculture in the Soviet Union, which had been originally made to finance the industrial sector of the economy, seems to have helped it to build up a powerful heavy industry. But it remains to be proved whether such a sacrifice was necessary. In any case it was paid for dearly by the Soviet People who went through severe hardships. Such situation is now practically impossible in other developing countries of the world. Moreover, the starting-off level of agriculture in the developing countries is much lower than it was in the Soviet Union at the beginning of its First Five Year Plan. Further, the history of industrial development shows that it has always been a slow process. As W.W. Rostow rightly points out that industrial development is much slower in getting under way when it has a little or practically nothing to start on.

Agricultural Transformation

A second group of writers like T.W. Schultz, W.W. Rostow and S. Kuznets argue that transformation of agriculture is a pre-condition for the development of the economy.[6] The argument of this group is based on the fact that if agricultural productivity was rising fast enough, the savings of the farmers could become available to finance investment in other sectors. Prof. Kuznets conclusions are explicit on this point. According to him "it is a pre-condition of industrialisation as a world-wide phenomenon that productivity of labour increases sufficiently to feed, at higher per capita levels, a larger proportion of the labour force than could be fed before. And as estimates have shown, in most of the developed countries, product per worker in agricultural sector increased more than the product per

---

6. T.W. Schultz. Transforming Traditional Agriculture, New Haven, Yale University Press, 1966 and W.W. Rostow, Stages of Economic Growth, Cambridge University Press, Cambridge, 1960.

worker in the rest of the economy combined. At the danger of stressing the obvious, one may claim that an agricultural revolution—a marked rise in productivity per worker in agriculture—is a pre-condition of the industrial revolution for many sizeable regions in the world."[7] Professor Rostow also observes that "A good part of the working capital for modernized industry must come from rapid increases in output achieved by higher productivity in agriculture."[8]

Historical records clearly show that no country has moved from stagnation into the take-off stage of economic development without first achieving a substantial gain in agricultural productivity. The United Kingdom, Germany, Japan and a few other countries have relied heavily upon imports of agricultural products. But this was usually after they had achieved substantial economic progress in agriculture and after industrial development was well under way. In a few of the less advanced countries, development of petroleum, mineral and other natural resources has led to large increases in income for a small part of the total population. But where this natural resources development has not been accompanied by improvement in agricultural productivity, most people have continued to live under conditions of abject poverty.[9]

Economic growth of the less advanced countries, therefore, depends heavily upon improving the performance of the agricultural sector just as it did in the more advanced countries in the early stages of their development. In less advanced countries agriculture accounts for 60 to 70 per cent of the total employment and for 40 to 50 per cent of the gross national product. But similar conditions prevailed in the advanced countries only 150 years ago. More recently, they have prevai-

---

7. Simon Kuznets, Lectures on Economic Growth, John Hopkins Press, Illinois, 1954, p. 63.
8. W.W. Rostow, Stages of Economic Growth, *op. cit.*, p. 83.
9. Raymound P. Christersen, Agricultural Progress in Less Developed Countries in Economic Development, published in "The Modernisation of Farming", Iowa State University Press, Iowa, USA 1965. p. 25.

led in Japan and other countries where economic growth has been rapid in recent years.[10]

Walking on 'Both Legs'

A third group of writers like Ragnar Nurkse and Arthur W. Lewis suggest a balanced growth approach giving equal priority to both the farm and non-farm sectors for healthy economic progress.[11] The philosophy of balanced growth envisages the simultaneous development of agriculture and industry so that industry may supply the basic needs in the form of inputs to agriculture which in turn may supply its surpluses at low prices for the expansion of the industrial sector. Arthur W. Lewis explains the idea as follows :

"Suppose that there is considerable innovation in the agricultural sector producing food for the home market. The result is either surplus food to sell to the towns or a surplus of labour in agriculture seeking non-agricultural employment, or some combination of both. If manufacturing industry is growing simultaneously and at the right rate, it can absorb both the surplus of goods and the surplus of labour. If it is not, the terms of trade will move against agriculture, and as there will be a surplus of farm-labour as well as farm-products, agricultural income will be depressed, and further investment and innovation in this sector may be discouraged . . . . Exactly the same difficulties arise if economic development is concentrated upon industrialisation, neglecting the agriculture, as happened in the U.S.S.R. There is, then, an acute shortage of agricultural products and an inflation of their prices, which drives up all other prices in a spiral movement. There is also difficulty in disposing of the manufactures at a profit...If agriculture stagnates, the capitalist sector cannot grow, capitalist profits remain a small part of the national income and savings and investment which are correspondingly small. Smooth

10. *Ibid.*, p. 26.

11. Ragnar Nurkse, Problems of Capital Formation in Underdeveloped Countries, Oxford University Press, Oxford, 1958 and Arthur W. Lewis, The Theory of Economic Growth, George Allen and Unwin Ltd., London, 1961.

economic development requires that industry and agriculture should grow together."[12]

The balanced growth approach has been successfully implemented by the United States as a goal of its economic policy and has resulted in significant improvement in its per capita income in every sector of the economy. For its agricultural base, the United States was obviously most favourably situated. Initially, underpopulated and having an abundance of land, it was able to produce large food and fibre surpluses and, at the same time, by the direct beneficiary of the Industrial Revolution which created a ready-made demand and rapidly-growing markets for its agricultural products.[13] Based on the experience of the United States, this group of writers points out that a developing country must achieve advances simultaneously over a broad range of activities. Further, they point out that the overall transformation of society in the social, cultural and economic spheres, can be sought through proper development of the infrastructure in the early stages. Many of the developing countries which have implemented and are implementing planned economic development have ignored the importance of the infrastructure in the process of economic development. The neglect of this important factor at the early stages of development is one of the main reasons for the slow rate of growth of some of the Middle East countries.

It is, of course, true that countries should avoid any dogmatic application of principles and theories which were evolved in the nineteenth century in Europe and other countries under different conditions from those prevailing in the underdeveloped countries. But under the present set-up of these countries it would be inadvisable to grant absolute

12. Arthur W. Lewis, The Theory of Economic Growth, *op. cit.*, pp. 276-77.

13. William H. Nicholis, The Place of Agriculture in Economic Development, published in Agriculture in Economic Development, Edited by Carl Eicher and Lawrance Witt, McGraw-Hill Book Company, New York, 1964, p. 17,

priority to any single sector of the economy.[14] Industrialisation, if achieved on an economically and technically sound basis can certainly bring about much more rapid increases in the productivity of the economy. Moreover, the overall progress of the rural economy requires modern means of production which will have to be provided by the industries of developing countries since they cannot make unlimited imports in view of their continuing unfavourable balance. Also, a substantial and rapid development of industry is necessary to raise the average value of exports and to achieve necessary reduction in the import of consumer goods. This principle appears to be indisputable. The recent history of industrial development in these countries, of course, provides us with practical confirmation of the great difficulties encountered. But efforts towards industrialisation must be continued and intensified, since without it an economy remains too dependent and subordinate.

Nevertheless it would certainly be a great mistake to try to bring about a rapid development of the economy solely by accelerating industrialisation without any far-reaching modernisation of agriculture. This might first slow down the overall development and ultimately even threaten the very success of such development. Agricultural modernisation and industrialisation are closely linked and all the developing countries should make a balance between them. It appears relatively easy to reconcile the two and 'walk on both legs' as the Chinese say.[15]

A Case Study of Ethiopia

Ethiopia is characterised by an economy that is still at a very low level of development and, consequently, a low level of real income per head. A closer comparison, based on the selected economic criteria, seems to indicate that Ethiopia, as far as development goes, falls among the less advanced countries of the world and indeed even of Africa itself. If it

14. U.N. African Agricultural Development, *op. cit.*, p. 83.

15. *Ibid.*, p. 17.

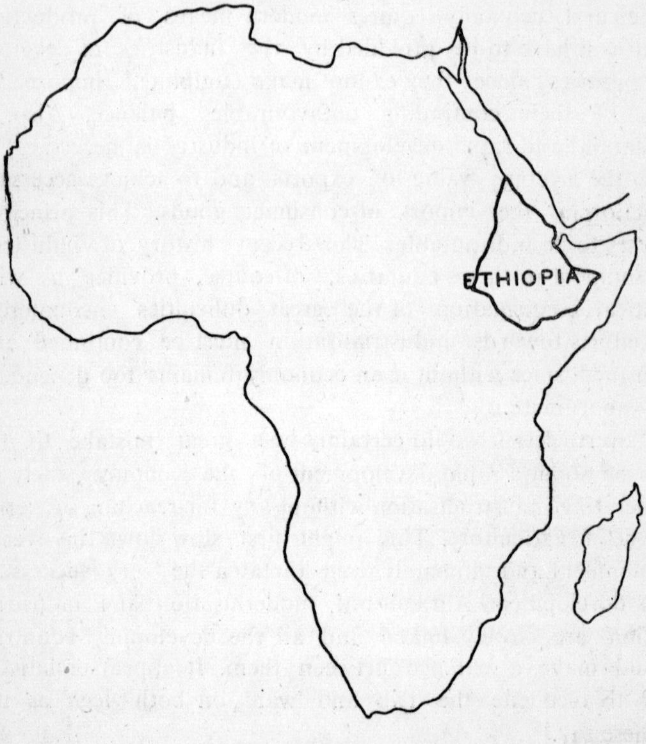

Fig. 1. Ethiopia's Location in Relation to the African Continent.

is judged by the GDP per capita, a population of 41 million has a GDP per capita of US $110.[16] The percentage of its active population engaged in agriculture is among the highest in both Africa and the rest of the world, but as in most less developed countries, the relative contribution of agriculture

16, The World Bank, Financing Adjustment with Growth in Sub-Saharan Africa, 1986-90, Washington, 1986, p. 67.

to the Gross Domestic Product is considerably lower. Statistically, while about 80 per cent of the total active population is engaged in agricultural production, the contribution of agriculture to the GDP seems to be about 50 per cent. "This is unmistakably an indication of extremely low productivity of agriculture in comparison with other countries under the conditions of backwardness."[17]

The position of Ethiopia in comparison to some of the countries of Africa seems to be more closely indicated by the high relative share of subsistence production. The share of this sector in the GDP has been estimated at about 40 per cent, while in other countries such as Ghana, Nigeria and Cameroun it does not exceed 33 to 35 per cent. In Congo and other mining areas the share of subsistence production goes as low as 12 to 9 per cent. The relative shares of mining and manufacturing in Ethiopia's GDP are small. Since about 80 per cent of the population earns its livelihood in low productive agriculture and no substantial change of this structure can be realistically expected in the near future, the per capita income of the total population will greatly depend on the per capita income of the agricultural population.[18] Furthermore, the low income of the agricultural population will also keep the level of wages in other branches low and thus their productivity too. Not only for the reasons of justice, but to minimise the problem just mentioned above, a policy of increasing productivity and income of the agricultural population must be planned and implemented.

Food Supply

At the time of implementation of planning, food production and consumption levels in Ethiopia, though relatively low both in calories and protective nutrients, could be considered

---

17. Ministry of Planning and Development, The Position of the Ethiopian Economy in Africa, Documentation No. Eco 3, pp. 1-2.
18. Ministry of Planning and Development, Third Five Year Development Plan (Second Draft) 1961-65 EC, Agriculture, Forestry, Fishery and Water Resources, Addis Ababa, 1968, p. 7.

TABLE 1.1

Population and GNP Per Capita of Low-Income Countries in Sub-Saharan Africa

Countries	Population (Million) Mid-1984	GNP Per Capita
Low-income Countries	260	220
1. Ethiopia	42.0	110
2. Mali	7.3	140
3. Zaire	30.6	150
4. Burkina	6.6	160
5. Guinea-Bissau	0.9	180
6. Niger	6.3	190
7. Malawi	6.8	210
8. Tanzania	21.5	210
9. Burundi	4.6	220
10. Uganda	14.3	230
11. Togo	2.9	250
12. Gambia	0.7	260
13. Somali	5.2	260
14. Benin	3.9	270
15. Central African Rep.	2.5	270
16. Madagascar	9.7	270
17. Rwanda	5.9	270
18. Guinea	5.9	300
19. Kenya	19.7	300
20. Sierra Leone	3.7	300
21. Sudan	21.5	340
22. Ghana	13.4	350
23. Senegal	6.4	380
24. Chad	4.9	...
25. Mozambique	13.4	...

Source : The World Bank, Financing Adjustment with Growth in Sub-Saharan Africa, 1986-90, Washington, 1986, p. 67,

to have been roughly satisfactory.[19] The growth of agricultural output was sufficient to meet the local demands of the population. But such growth could not meet new demands of a developing country like Ethiopia in the subsequent years where modern development was taking place. That brought forth changes in the growth of demand for food by increasing both the rate of growth of the non-agricultural population and of income and consequently consumption per capita. With the economic development in the period to come, these changes will further accelerate and the total demand for food in general and of higher quality food in particular will increase faster than the rate of growth of population. At present there are obvious signs that food production is not keeping pace with the demand. Recently, there has been almost chronic shortage of food supply in some of the areas and consequently occurrence of famine. In order to avoid economic problems, it is essential that food growth be steadily increased at a faster rate than the growth of population.

In a developing country industrialisation must rely significantly on raw materials from agriculture as other sources of raw materials are not available in sufficient quantities. Moreover, industries using agricultural law materials require relatively low capital and modest skilled labour. Thanks to climatic conditions, Ethiopia already grows or can grow a wide range of different agricultural raw materials to meet the needs of potential industries.[20] Many industries in existence, however, have operated at much less than their capacities because of inadequate supply of agricultural raw materials. Some agricultural products in the form of raw materials which could be produced in Ethiopia are still unnecessarily imported particularly cotton and coarse fibres. Much higher production and improved marketing of agricultural raw materials are needed to stimulate the growth of the industries.

---

19. *Ibid.*, p. 2.
20. Ministry of Planning and Development, Economic Development in Ethiopia and Prospective Trade and Economic Relations with Great Britain, Documentation No. Eco. 2 p. 4.

Export Items

The plan for a substantial acceleration of economic and social development depends upon the increase of exports in order to permit sufficient imports necessary to achieve the desired goals. Over 90 per cent of present exports are from agriculture or of products derived therefrom and there is a little hope that an appreciable change in the composition of exports may take place during the coming years.[21] Coffee, the main item, constitutes about two-thirds of the total exports. The production of other export items such as oilseeds, pulses and animal products must be increased. As local consumption of most of these items will also increase, it is important to concentrate most of the efforts to quickly increase their production. Further, a dollar of foreign exchange saved by import substitution can be as significant as a dollar earned from export provided local production costs are not far above the import prices. Cotton fibres, coarse fibres for sack and bag cloth and wheat are some of the items which offer some of the best opportunities for immediate development.

The agricultural sector is also important as a market for the goods of manufacturing industries, as a supplier of labour to other sectors and as a contributor to the national income and savings of the country. In a subsistence sector, the purchasing power of agricultural population is very low and the consumption of purchased goods is thus very limited. The narrow domestic market can absorb only small quantities of manufactured goods. In a country like Ethiopia with its relatively small population such a market cannot stimulate the development of various manufacturing industries. This type of situation also limits its capacity for economic operation. Ethiopian agriculture uses negligible amounts of fertilisers, pesticides and other chemicals and very small quantities of agricultural implements and machines. The negligible demand of these items does not justify the establishment of the corresponding industries in the country and as such restricts the desirable growth. A more rapid change

21, Ministry of Planning and Development, Third Five Year Development Plan, (Second Draft), *op. cit.* p. 3,

in the structure of economy is not possible if agricultural income is allowed to remain low.[22]

Further more, in countries with a predominantly agricultural population like Ethiopia, the main supplier of additional manpower to other sectors is agriculture. If there are considerable innovations in the agricultural sector, the result is a surplus of labour force seeking non-agricultural employment. If manufacturing industry is also expanding rapidly, it can utilise this surplus labour-force.

Five Year Development Plans

Ethiopia is the first African country to have begun a comprehensive planning for its economic and social development. By adopting planning as a method of formulating an economic policy, it showed its resolution to accelerate the development process and shorten the time necessary for raising the economy and society to a higher level.[23]

Before launching the First Five Year Plan, there were separate programmes and plans for different sectors in the country. During 1944-45 the Government initiated a Ten-Year Programme for industrial development which could not be carried out fully as prevalent conditions did not permit its completion.[24] In the sphere of agriculture and animal husbandry steps were taken to increase production as well as to carry out research. During 1952-53 improvement of roads and its extension work was carried out on the basis of a four year progammc. Various studies were conducted on mineral and water resources. The total investment on an average was about Eth. $ 80 million per year during this period. Of this, net investment outside agriculture amounted to Eth. $ 50 million. In agriculture, with the exception of small credit from the Bank, investments were mainly non-monetary in character and appeared in the form of increase in livestock, construction of houses and tools and implements

22. *Ibid.*, p. 5.
23. Ministry of Planning and Development, Evaluation Report on the Economic and Social Development during the First Five Year Plan, Documentation No. Eco. 6, p. 1.
24. Imperial Ethiopian Government, Five Year Development Plan, 1957-61, Addis Ababa, p. 1.

made by agricultural producers. In the total annual volume of investment the share of investment on agriculture amounted to 1.7 per cent. Since agriculture contributes the bulk of the GDP, slow growth of agricultural production was the main reason why the rate of growth in the net national product was relatively small.

During the First Five Year Development Plan period (1957-61) it was realised that agricultural potentials had not been sufficiently utilised and that its development could provide a better and more varied diet for the population, more raw materials for manufacturing industries and large surpluses for exports.[25] But at the same time it was also realised that one of the principal obstacles to more intensive economic development of the country under existing circumstances was the rudimentary infrastructure. It was, therefore, necessary to give high priority to these basic facilities. Out of the total monetary investment, about two-thirds were allocated for infrastructure investment.

The reason for the development of the transport and communication system was that vast and remote areas often having potential for the economic deveolpment were isolated. The high transportation costs were unduly raising the prices of many commodities thereby obstructing the expansion of production, consumption and export. The investmment in this sector was 45 per cent of the total monetary investment for the five years. In determining the priority, an important place was also given to agriculture. It was decided to increase the monetary investment in agriculture though it constituted only a small portion of the proposed investment. "It must, however, be remembered that in previous years monetary investment which is far more productive than investment in kind, was almost significant and that a significant increase in agricultural production is possible under existing conditions without large monetary investment."[26] In addition to the rural investment in kind estimated at Eth. $130 million, monetary investment in

25. *Ibid.*, p. 19.
26. *Ibid.*, p. 35.

agriculture was about Eth. $ 40 million or 7.5 per cent of the total monetary investment for the five years.

The First Plan had given priority to infrastructure, the Second Plan gave greater emphasis to directly productive undertakings, particularly manufacturing industry, mining, electricity, etc., Out of the total monetary investment, the industrial sector got a share of 31 per cent. Transport received 19.6 per cent of the total monetary investment. The agricultural sector was also given a reasonable share of the total investment. An amount of Eth. $ 335 million was allocated to this sector. It was also expected that farmers' contribution in labour would be Eth. $ 113 million. So the monetary investment for agriculture was Eth. $ 242 million or 16.6 per cent of the total monetary investment for the five years (1963-67) although about 42 per cent of the total monetary investment was spent during the Second Plan period.

Agricultural Plan

It was in the Third Five Year Plan, (1968-1973) that there was a substantial shift of emphasis from the preceding plans by giving high priority to agriculture.[27] While formulating the Third Plan it was emphasised that the expansion of non-agricultural activities "will generate employment and will result in monetary demand for agricultural products ; but if this demand is not accompanied by increased productivity in agriculture, the price of food will inevitably increase"[28] Such a development would hamper the expansion of non-agricultural activities, due to failure to generate the necessary additional demand, within the agricultural sector, for non-agricultural products.

The Third Plan required a total investment of Eth. $ 3,415 million, of which approximately Eth. $ 550 million was estimated in kind and rest (Eth. $ 2,865 million) was monetary investment. This investment was in gross terms. Out of the monetary investment, transport and communications got 21.8 per cent. Next comes housing and manufacturing industry which were allocated 18.3 per cent and 18 per cent respectively.

27. Imperial Ethiopian Government, Third Five Year Development Plan 1968-73, Addis Ababa, p. 19.
28. *Ibid.*, p. 204.

Total monetary investment allocated to agricultural sector was Eth. $ 312 million (Eth. $ 7 million for forestry and Eth . $ 5 million for fishing) or 10.9 per cent of the total monetary investment. The total investment both monetary and non-monetary on agriculture for the planned period was Eth. $ 850 million, out of Eth. $ 3,415 million. In percentage, it was about 24.9 per cent of the total investment. It is in this respect that the Third Plan was considered to be an agriculture-oriented plan.

TABLE I.2

Monetary Investment During Second and Third Plans

(In million)

Sector	Second Plan	Third Plan
Agriculture	242.2	312.1
Manufacturing Industry*	318.0	565.0
Mining	74.0	179.4
Electrical Power	58.0	120.0
Transport & Communications	335.0	624.0
Education and Training	22.1	108 6
Health	39.0	41.0
Housing	242.0	524.6
Others	248.9	390.0
Total	1451.2	2,865.0

* Includes handicraft and small industries.

Source : Imperial Ethiopian Government, Third Five Year Development Plan, (1968-73) Statistical Appendix, Part I & II.

Socialism and Agriculture

In September 1974, the armed forces dethroned the Emperor and appointed their own Chief to lead the nation. The Military Government announced the policy of Ethiopian Socialism based on social and economic equality. In a policy statement read over the Radio, the Government said that it would reconstruct the economy with the motto of "Ethiopia First" which means "Ethiopian Socialism." The exploitation of man would be banned and the structure of the future government would be based on the right of the people to administer themselves. The Government also announced the inauguration of Ethiopia's first rural upliftment programme to explain the philosophy of Ethiopia First and the concept of self-reliance and co-operation, to convince the people that every one has to bear an equal share of the responsibility for the progress and development of the country, and to impress for the dignity of workers on the minds of the people. It also outlined three basic reforms, i.e., nationalisation of basic and key industries, land reform and tax reforms designed to promote an equitable distribution of resources.

Ever since the inception of the Revolution the task that demanded most of the time and energy of the Government is the endeavour to accelerate the growth of agricultural sector and lay down the necessary pre-conditions for building a socialist society. "The accelerated growth or stagnation of this sector with its considerable contribution to production and employment will determine the nation's general economic growth or stagnation. There can be no choice at this initial stage but to give priority to the development of this economic sector in the overall endeavour to accelerate the general economic growth of the country."[29] The importance of agriculture can be seen in its increasing share of the capital expenditure since the Revolution. In 1973-74, agriculture was allocated about 22 per cent of the capital expenditure. Because of the drought rehabilitation activities, the share of this sector rose to 33 per cent of the capital expenditure in 1974-75. Agriculture was

29. Documents and Resolution of the 2nd COPWE Congress, Addis Ababa, January, 1983, p. 38.

allotted 30 per cent of the capital budget in 1975-76. As a result of land reform, this sector took up the largest portion i.e., 41 per cent of the capital expenditure in 1976-77.[30]

In continuation of these efforts, the Government launched in 1978 a development campaign, the Zemecha which placed emphasis on the reconstruction and restructuring of the economy. The short term aim was to eliminate illiteracy and unemployment while at the same time meeting consumers demand. The long term aim was to establish a diversified economy with agriculture providing the base and industry playing a leading role. "Our objective is to affect a radical change in our general national economic structure in building socialism, there is no choice but to base our economy on farm and small scale industries."[31] The expansion of agriculture is required to supply raw materials to processing industries, cash crops for export and to alleviate food shortages. The development campaign aimed to increase agricultural production by 4 per cent and 8 per cent during the two annual Plan years.

In 1980, a long term economic development plan called Ten Year Investment Programme with a budget of 26,562 million Birr was launched. Of which 39 per cent was allocated, for the first half of the decade 1980-81 to 1989-90 with the balance for the second half. The most important target of the programme is to double the real Gross Domestic Product from 8.4 million Birr to 17.3 million Birr. This would be possible with the 7.5 per cent annual rate of growth of real GDP. Agricultural target is to increase its productivity by 60 per cent and for this about 12.4 per cent of the total investment cost has been allocated as against 13.6 per cent to manufacturing industry and 29.7 per cent to transport and communications.

As far as agriculture is concerned, the programme sets out a strategy to "increase the productivity of peasant agriculture through better production techniques, use of fertilisers, impro-

30. Seleshi Sisaya and Eile Stommes, Agricultural Development in Ethiopia : Government Budgeting and Development Assistance in the Pre and Post 1975 Periods, Journal of Development Studies, Vol. 16, No. 2, January, 1980, p. 169.
31. African Research Bulletin, Economic, Financial and Technical Series Vol. 16. No. 1, p. 4884.

TABLE 1.3
Estimated Investment of Ten-Year Development Plan (Million Birr)

Sector	Total Investment	As Percentage of Total Investment
Transport and Communications	7890	29.7
Relief and Rehabilitation	4342	16.3
Industry	3625	13.6
Agriculture	3285	12.4
Energy and Water Resources	3167	11.9
Education	2320	8.7
Mining	976	3.8
Foreign Trade and Tourism	670	2.6
Health	300	1.1
Total	26,562	100.0

Source : United Nations Conference of the Least Developed Countries, Addis Ababa, 1981.

ved seeds...and by increasing farm labour productivity through co-operative ventures."[32] It also opts for employing labour-intensive techniques of production in agriculture and rural road construction and the expansion of irrigated farming which would permit the realisation of dependable and adequate food supplies and enhance the country's export capabilities.

The above analysis clearly shows that the importance of agricultural development for the overall economic progress has been recognised in Ethiopia. Efforts have been made to develop this sector both during the pre and post Revolutionary periods. But the available figures show declining rates of

32. Perter Schwab, *Ethiopia—Politics, Economics, and Society*, Frances Printer (Publishers), London, 1985, p. 78.

growth in agricultural output during the last two decades. In 1973 Ethiopia underwent a prolonged period of drought and famine that took the lives of hundreds of thousands of the people. Similarly in early 1984, as reports indicate, hunger again invaded Ethiopia with a severity which approached that of 1973. With the severe reduction of agricultural output, it is likely that another such tragedy might occur unless the agricultural development programme is implemented seriously.[33]

33. *Ibid.,*. p. 80.

CHAPTER TWO

Socio-Economic Characteristics

ETHIOPIA lies between a latitude of 3 degrees and 18 degrees North and a longitude of 33 degrees and 48 degrees East. It is a roughly triangular area of about 1,221,900 square kilometres with a coastline of about 1,200 kilometres.[1] It faces the Arabian Peninsular across the Red Sea and is situated in the North of the Equator. It has a common border on the west and north-west with Sudan, on the east and south-east with Somalia and south with Kenya. Close to the Red Sea coast are a number of island groups most important of which is known as the Dahlak Archipelage. Since it is situated in north-east Africa it is usually included, together with Somalia and Djibouti, in the region known as the Horn of Africa.

Before September 1974, Ethiopia was a constitutional monarchy with the Emperor as its head of State. The Chamber of

1. The Technical Agency, IEG ; The Resources and Economy of Ethiopia, Report No. 13, Stanford Research Institute California, Feb., 1969, p. 9.

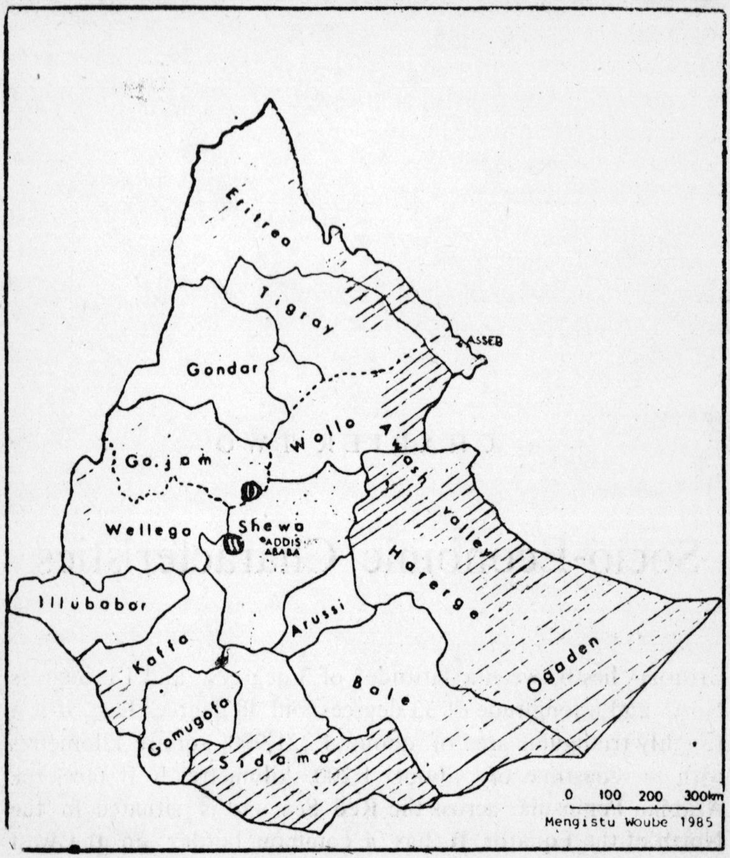

Fig. 2.2 : Provincial Divisions

Deputies were elected once in four years on the basis of universal adult suffrage. The Revised Constitution of 1955 formed the supreme law of the country and guaranteed equal protection before the law to individuals and safeguarded to their live, liberty and property. In this way the Constitution gave Ethiopia to some extent a modern form of Government.[2] From the administrative point of view, Ethiopia was divided into 14

2. H.P. Huffnagel, Agriculture.in Ethiopia, RAO/U.N., Rome, 1961, p. 5,

provinces—sometimes called Governorate-General—which were sub-divided further into 'Awraja' or sub-provinces. A Governorate-General was headed by a Governor-General (or a representative of His Imperial Majesty). He was responsible for the maintenance of public security, proper administration and dispensation of justice. The Governorate-General Council advised him on matters relating to the welfare, the advancement and the prosperity of the inhabitants. Council was composed of senior officers of the province. In 1976, they were officially renamed as administrative regions. Smaller administrative units are the sub-provinces (awraja), the district (woreda) and the kebele (locality, urban district).

Climate

Although the entire area of the country lies within the tropical zone because of its proximity to the Equator, it does not experience a tropical climate.[3] Owing to Ethiopia's generally high elevation, its climate is temperate. In other words, altitude is an important factor which determines the temperature that prevails in Ethiopia. Taking the influences of altitude and climate, Ethiopia is broadly distinguished by four climatic zones:[4]

(*a*) a mountain zone called 'Wirch' comprising about 3,800 metres where a high rainfall over several months precludes cultivation. Its annual average temperature is 14°C with alpine conditions including hailstorms ;

(*b*) a cool zone called 'Dega' with average temperature from 10°C to 18°C and altitude between 2,400 to 3,800 metres where the elevation affords conditions for the cultivation of cool temperate cereals like wheat, teff and barley ;

(*c*) a temperate zone called 'Weina Dega' with average temperature between 16°C and 20°C and altitude

3. U.S.A. Operation Mission to Ethiopia, Point 4—The Agriculture of Ethiopia, Vol. I, January, 1954, Reprinted in October, 1957, p. 149.
4. International Bank for Reconstruction and Development/IDA ; The Economy of Ethiopia, Vol. 2, Agriculture, Africa Deptt. August, 1967, p. 1.

between 1,800 and 2,400 metres where in addition to the crops grown in the cool zone, maize, sorghum and millet come into prominence. In the south-west, rainfall is high and hence coffee and ensete plantations are also found ;

(*d*) a hot zone called 'Kolla' below 1,800 metres and with an average temperature about 20°C and low rainfall where there is no cultivation except under irrigation. Here tropical and semi-tropical crops are grown.

It may also be mentioned here that at elevations above 2,000 metres, high daily variations in temperature occur between November and January. This factor to some extent adversely affects the introduction of improved varieties of crops.[5] But less variations are observed in daily temperature at lower elevations. Therefore such areas are often ideal for agriculture and crops can be grown all around the year. But it must be remembered that the rainfall tends to decrease at lower altitudes with very limited precipitation in contrast to the highlands where agriculture is presently more developed and the amount of rainfall is generally satisfactory.[6]

In general, Ethiopia's rainfall is adequate for agriculture, except in the semi-arid and desert regions. Although there are year to year variations in the country, there are two types of rainy seasons. They are known as the 'big rains' and the 'small rains.' The former normally occurs during July to September and the later usually falls during February, March and sometimes in April. "This climatic feature gives Ethiopia some comparative advantages in the production of certain crops particularly certain fruits and vegetables that can be marketed when the supply is short in the major world markets."[7]

Soil Fertility

Although the fertility of the Ethiopian soil varies from

5. The Technical Ageney, IEG, The Resources and Economy of Ethiopia, *op. cit.*, p. 15.
6. *Ibid.*, 15.
7. *Ibid.*, p. 16,

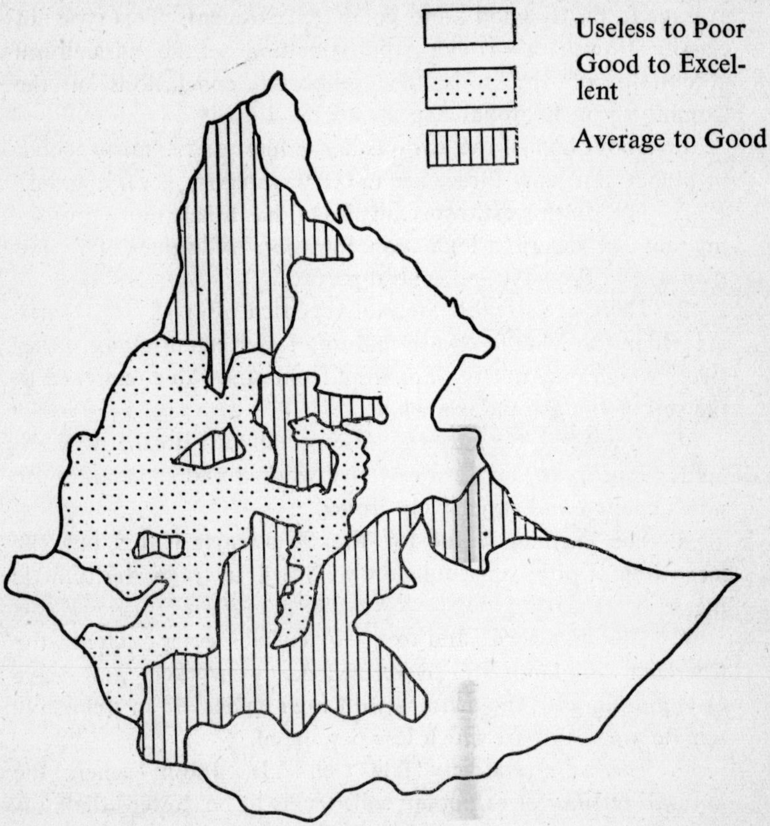

Fig. 2.3 : Classification of Ethiopian Soils

region to region, it is generally considered to be fertile. Two main types of soils are recognised in Ethiopia—the red to reddish-brown clayey loam and the black soil.[8] The former is usually of good fertility but their major deficiency in plant nutrients is phosphorus. The black soils have a tendency to dry out quickly and to crack badly thus hampering the cultivation. On the basis of a map prepared by the Committee on Regional Aspects of National Planning in Ethiopia, the fertility of the soils can be grouped into three categories—Useless to poor;

8. H.P. Huffnagel, Agriculture in Ethiopia, *op. cit.*, p. 37.

average to partly good ; and good to excellent. This type of classification gives a rough approximation of the agricultural potentiality of the soils. The important conclusions of the Committee on Regional Aspects are.[9]

1. About half of the awrajas has values of 7, 'fairly good' or higher, and only three are in the 9 category, 'very good.'

2. The south-west, especially Kefa, has much more promising soil ; in general a high concentration of values of 7 and over are in the west and central parts.

3. There are sizeable areas of very rich alluvial soil particularly in the Middle Awash and to a lesser extent along other rivers which may not be apparent from the value assigned to the soil of the awraja as a whole.

4. In general, the eastern and especially the northern highlands appear to be seriously overworked and some areas are now depleted and extensively eroded.

5. The median value for the 99 awrajas is 6.5 but the large area of poor soils pulls the weighted average (arithmetic mean) down to 5.

6. The northern and to a somewhat lesser extent, the Chercher highlands are overcrowded, overworked and have little margin left ; the awrajas with high values are in relatively remote areas and are much less populated.

7. There are relatively few rich soils, and in general the natural fertility of Ethiopian soils seems to be lower than has been assumed.

The above conclusions show that more detailed studies need to be undertaken particularly on yield. But available statistics clearly reveal that the bulk of the more heavily populated highlands are badly damaged and agricultural land is now relatively limited in area. Intensive development areas will have to be carefully chosen. There are of course relatively good areas in some of the crowded areas and preliminary crop cutting experiments show that many farms in Chilalo, Yeror and

---

9. Ministry of Planning and Development, Committee on Regional Aspect of National Planning in Ethiopia, Part I, Addis Ababa, August, 1967, p. 24.

Kereya show relatively high yield under the traditional practices.[10] But extensive and easily available areas of rich soil are not plentiful. Further, it must be noted that the major problems of Ethiopian soil particular in the highlands are erosion and thus Ethiopia loses a great area of fertile land every year.[11]

Land and its Utilisation

The total area of the country is 1,221,900 square kilometres. According to various estimates the distribution of the total area among different categories is given in Table II.1. It will be seen from the Table that the productive area is relatively high i.e., 76 per cent. The same holds with the agricultural land which is about 69 per cent of the total area of the country. But the area of arable land (crop land, fallow land and orchard) is relatively very low amounting about 10 per cent if a large part of the grass land is also considered to be suitable for cultivation. Crop land is estimated to occupy about 8.3 per

TABLE II.1
Land and its Utilisation

Category	Million Hect.	% Distribution
Agricultural Land	84.1	68.8
Crop Land	10.2	8.3
Fallow Land	2.2	1.6
Orchard, Vineyards & Coffee	0.7	0.6
Grass Land	66.0	54.0
Swamps	5.2	4.3
Forests	8.8	7.2
Others (Barren land, water etc.)	29.3	24.0
Total	122.2	100.0

Source: Ministry of Planning and Development, Third Five Year Development Plan (Second Draft) 1961-65 EC (1968-73), Addis Ababa, p. 18.

10. *Ibid.*, p. 25.
11. The Technical Agency, IEG, The Resources and Economy of Ethiopia, *op. cit.*, p. 15.

cent of the total area and that too is largely restricted to the highlands at altitudes between 1,700 and 2,700 metres having an adequate and favourable distribution of rainfall. Cropping is thus most developed in the central highlands, north of the Rift Valley and on the northern margins of the highlands of the Rift Valley. However, much cropping is also done under marginal rainfall conditions in the northern provinces.[12]

Most of the grazing land is associated with the cropped area in the highland regions developed from previously forested land. A portion of the grazing lands is grass fallow left to regain fertility under the natural grass cover. These grass lands within the cropped areas are heavily grazed. Increasing pressure of population and the necessity for more cereals has shortened grass fallow and with the result that the scope for extending the cultivated area is limited. Semi-desert extending from lower Awash Valley covers an extensive area and true desert conditions occur there. Another large semi-desert area occupies half of the provinces of Harerge and Bale. These desert areas are exploited by nomads who travel with their herds of cattle in search of grazing land.

So, within the central and northern highlands, there is limited scope for increasing cropped areas though in some regions of northern highlands the crop resources are already failing to meet the needs of the people and severe local food shortages have occurred. A limited extension of cropped area is possible and is occurring in the highlands of south of the Rift Valley. But the scope for further extension is very limited. The central highlands offer opportunities for bringing areas under cultivation and new settlers are already migrating into the unoccupied areas of Gojam and Wollo. But at present the means of communications are very poor and the tsetse-fly is a known risk at the lower elevation.[13]

Farming Population

According to the "Office of the Population and Housing

---

12. International Bank for Reconstruction and Development, The Economy of Ethiopia, Vol. II, Agriculture, *op. cit.*, p. 2.
13. *Ibid.*, p. 2.

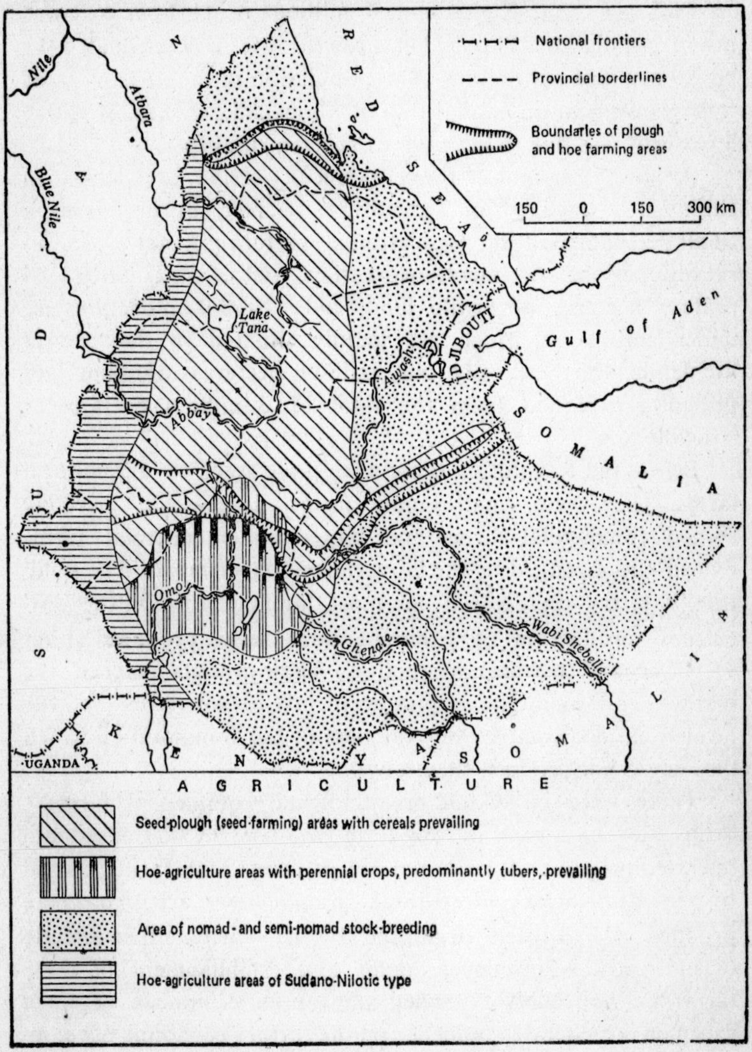

Fig. 2.4 : Principal Farming Areas.

Commission", the population of Ethiopia in 1984 was more than 42 million.[14] Of this, the rural population was 38 million

14. Office of the Population and Housing Commission, Population and Housing Census Preliminary Reports, Vol. I, No. I, 1984, p. 14.

or 88.7 per cent. Remaining 4 million or 11.3 per cent was urban population. Population growth rate was estimated to be 2.9 per cent per annum. The population in the urban areas was growing at an annual rate of about 5.5 per cent, of which 3 per cent was due to net migration from rural to urban areas.

About 85 per cent of the economically active population in Ethiopia is engaged in agriculture. In fact, in rural areas, a small percentage of the population is performing non-agricultural occupations. As mentioned earlier, out of the country's 84 million hectares of fertile land, only 10 million hectares are under cultivation which are mainly situated in the central highlands. Further, only one million hectares of land are presently exploited from the total of 3 million hectares of irrigable land.

Before the Revolution there were three types of agricultural farms. They were small family farms, large-scale privately owned commercial farms and large-scale commercial farms owned and managed by the Government departments. The small family farms represented the most dominant form in the agricultural production of Ethiopia. Such farms produced about 95 per cent of the total production. Their product reached the market only when any surplus accidently remained or when the farmers needed some money or some of the commodities which they could not produce themselves.

There were large-scale private-owned commercial farms—farms run on a basis of long-term land tenancy (organised and operated mostly by foreign nations or firms) and farms owned by persons or firms and operated on their own accounts where production had been organised by the owners, either directly or indirectly.[15] The development and establishment of these farms had noticeably expanded and quite a number of them executed considerable influence on the further economic development of the country. All these farms were equipped with modern machinery and so well organised. They produced mostly key cash crops.

Such farms included the *Tendaho Plantation Share Company* in the Awash Valley. The decision in 1960 to initiate the

15. Ministry of Information, Agriculture in Ethiopia, *op. cit.*, p. 18.

cultivation of cotton on plantation scale marked the beginning of a new chapter in the field of cotton-growing in Ethiopia. The Tendaho Share Company was formed on 4th July, 1961. The capital of the company was subscribed 51 per cent by Mitchell, Cotts Groups and their associates and 49 per cent by Ethiopian nationals.[16] Cotton was planted over an area of 3,500 hectares in Dubte and 1,530 hectares at Dit Bahari making the total land under cultivation 5,030 hectares. The lint cotton produced by the company was used by the cotton textile factories in Ethiopia. A small quantity had been exported in order to establish the marketability of the cotton in overseas countries. The whole quantity of the produced cotton seeds was exported.[17]

The Cotton Plantation in Tessenei (Eritrea) was one of the major agricultural farms in Ethiopia. It was situated near the Etho-Sudanese border. Though the idea was originated as far back as 1900 but the farm for all practical purposes could not attain its aim. Even during the British Administration of Eritrea the farm was almost non-existent and this farm was converted for food production due to the shortage of food in that region. However, the cultivation of cotton was again permitted in 1949. This concession which belonged to SIA Ltd. was taken over by the Etho-Italian Share Company in 1965. The climate of the area for the production of cotton and the possibilities to use the Gash river for irrigation had made it an ideal place for cotton cultivation. Before 1974, a total of 8,000 hectares were brought under cotton cultivation. Apart from cotton, the management also cultivated 2,000 hectares with durra.

Three sugar-cane plantations on most modern lines had been established in the Awash Basin. They were the *Wonji, Shoa and Metahara Sugar Estates*. The first two were the main sources of sugar supply in Ethiopia and the third was in its initial stage. The Wonji and Shoa Estates were situated in the

16. *Ibid.*, p. 19.
17. K.A. Cherian, Ethiopia-Today. An up-to-date Illustrated Review of Economic Development, Centre Planning Perss, Addis Ababa, June, 1969, p. 44.

plain of Nazaret, near Addis Ababa. In June 1951, an agreement was signed between the Imperial Ethiopian Government and the Dutch Concern HVA Ltd. by which the latter was granted a concession of 5,000 hectares for building up a sugar-estate and factory. In 1954 the sugar factory was opened and during the first milling session produced 16,000 tons of sugar. But during the subsequent years there was great expansion and in 1959-60 the factory was able to produce 40,000 tons of sugar. Towards the end of 1958 the Dutch holdings of HVA Ltd. were transfered to an Ethiopian Share Company, HVA-Ethiopia. This Company started the second sugar factory named Shoa. Since more land was needed, the Ethiopian Government granted a concession of another 1,600 hectares. The Shoa factory was opened in 1962. The total production capacity of the Wonji and Shoa factories totalled an average of 70,000 tons per annum. HVA-Ethiopia had initiated another sugar estate project known as HVA-Metahara. The cost of this project was estimated to be about Eth. $ 56 million.

In *Elaberet*, Eritrea, there was a modern agricultural farm for the production of fruits and vegetables and for raising cattle to supply milk for sale in Ethiopia and for export.[18] The estate consisted of 1,000 hectares of land. A large area was used for the cultivation of citrus, fruits and vegetables tomatoes and papaya. Of the 1,000 hectares, more than 500 hectares had been brought under cultivation. In the Awash Valley there were some fruit farms. Two of the important ones in this region were the Erer Gota farm and the Awara Melka. In both of them commercial types of fruits were cultivated. There were also small agricultural farms operated by Ethiopian nationals in different parts of the country.

There were also large-scale commercial farms in Ethiopia which were operated and managed by the State. The Ministry of National Community Development was the investor in these farms.[19] The funds for their establishment were allocated from the budget or derived from foreign loans (spent mostly for the purchase of up-to-date equipment, servicing workshops,

18. Ministry of Information, Agriculture in Ethiopia, *op, cit.*, p. 22.
19. *Ibid.*, p. 25.

etc.). Before 1974 three such farms had been established and were in the stage of further development. One was in Awassa where about 4,000 hectares had been made arable so far on cleared land. The farm was primarily oriented towards the production of oil-seeds and sisal. The other was in Botter where about 700 hectares of cleared land had been made available. Because of the rich and fertile pasture, this farm was mainly oriented towards the fattening of livestock for the slaughter-house industry. In the Arba-Minch region in the south, suitable for cotton production, the Ministry of National Community Development started a project and established a settlement from the neighbouring highland inhabitants. These farms will be discussed in detail in Chapter VII.

After the land reform proclamation, the classification of agricultural farms has changed. They are now called peasant farms, co-operative farms and State farms. The crop production of peasant farms still constitutes about 94 per cent of the country's overall crop production and 97 per cent of the total cultivated area. But due to under-development of the methods of production, the productivity of these farms is low. Ox-power is predominant on these farms particularly in the highlands. Usually deep ploughing is not possible with oxen. The soil surface is often only scratched which does not permit good water penetration or provide favourable conditions for plant growth. Fertilisers and insecticides are generally not used by farmers.

Co-operatives are still at their initial stages and co-operative farms produce only less than one per cent of the total country's cereal production. However, the future policy of the Government is to establish strong co-operative farms throughout the country by converting existing peasant associations into producers' co-operatives.[20] In 1975, the privately owned commercial farms were converted into State farms. Such farms account for about 1.5 per cent of the country's total land under cultivation and produce about 5 per cent of the total crop production. The State farms produce both the crops for local as well as for export purposes.

---

20. National Bank of Ethiopia, Annual Report, Addis Ababa, 1979, p. 11.

Water Resources

Few of the Ethiopian rivers are perennial and in size they vary greatly. Usually both the perennial and seasonal streams have cut deeply into the soil leaving small possibilities for irrigation.[21] Further, the use of their water for irrigation purposes is also greatly impeded by their difficult access. Nevertheless, a potential exists for the irrigation of relatively large areas from some rivers and lakes. The main Ethiopian rivers are :

TABLE II.2

Main Ethiopian Rivers

Name	Length in Kms.		
	Total	Inside Ethiopia	Outside Ethiopia
Abay	1,450	800	650
Wabi Shebele	1,130	1,000	130
Genale	850	480	370
Awash	1,200	1,200	—
Omo	760	760	—
Tekeze	608	608	—
Mereb	440	440	—
Baro	227	227	—
Angereb	220	220	—

Source : Central Statistical Office, Ethiopia—Statistical Abstract, 1980 Addis Ababa.

The Abay or Blue Nile is certainly the most important river of Ethiopia. It is more than 1,450 kms. long from its source in the Gojam highlands in Ethiopia to Khartoum in Sudan where it joins the White Nile. Of this, about 800 kms. are within Ethiopia. Many of its tributaries are perennial. The basin of the Blue Nile in Ethiopia is ideal for

21. Ministry of Planning and Development, Third Five Year Development Plan (Second Draft) *op. cit.*, p. 21.

agricultural production. Irrigable area in this basin comprises 550,000 hectares and about 557,000 hectares are estimated to be good for cattle raising and dry farming.[22]

The Awash river begins in the highlands southwest of Addis Ababa, around which it flows in a wide-semi circle, and follows a north-eastern direction along the Rift Valley into the Danakil plain. Although it is, for most of its course, a wide perennial stream with high floods, it crosses a series of lagoons and then disappears into Lake Abbe. Its main tributaries are the Akaki and the Kassam. The soil in the basin of the river has been found to be suitable for the cultivation of cotton, oilseeds, sugarcane, rice and tropical and subtropical fruits. In the lower middle basin of the river there are as much as about 365,000 hectares of land which can also be used for agricultural production. The fertility of the soil is good.[23] In view of the great economic potentialities of the Awash river, the Government established the Awash Valley Authority in January, 1962 to administer and develop the natural resources of the region. The studies undertaken reveal the possibilities of the development of the region for agricultural activities. The full development of the basin had been tentatively estimated to cost Eth. $ 430 million.

The Wabi Shebele is another important Ethiopian watercourse which is about 1,000 kms. inside Ethiopia. It rises in Arussi, winds along the Gugu and Chercher mountains, turns south-eastwards, crosses the Somalia frontier and eventually reaches the Indian Ocean. The lower part of this region is poor in rainfall while the upper part has abundant rainfall that causes floods every year.[24] The stabilised part of the population of this area is engaged in agriculture and cattle raising. This sector employs about 99 per cent of the total population. The soil is very hard. Studies conducted so far reveal that about 200,000 hectares of land can be irrigated to cultivate one to three crops a year. Other important rivers are Omo, Tekeze and Baro. There are also some important lakes such as Abaya, Awash, and Tana which provide opportunities for the irrigation of land.

22. K.A. Cherian, Ethiopia—To-day, *op. cit.*, p. 40.
23. *Ibid.*, p. 40.
24. *Ibid.*, p. 41.

But due to high investment costs, irrigation will probably be found to be practicable only where climatic and soil conditions render it possible for the soil to be used almost all the year under cash crops with high yields and favourable prices.[25] The high cost of transportation is at present the limiting factor for many of the potential irrigation schemes to be built and used. Except for a few irrigation schemes such as Awash Valley where transportation and market conditions favour the organisation of capital-intensive agriculture, and for very small schemes with low investment costs, it is difficult to take up large scale irrigation projects.

Transportation

An agricultural enterprise in Ethiopia itself is a transport organisation.[26] Most of the agricultural products are bulky and transport costs are an important fraction of the total cost. In a subsistence agriculture with low yield and small farms, the quantities for transportation are relatively small and distances are short.

There are two railway lines in Ethiopia. One goes from the port of Djibouti by way of Dire Dawa to Addis Ababa and is 781 kms. long. The other goes from the port of Massawa by way of Asmara beyond Agardot and is 305 kms. long. Both railways are narrow gauge. Statistics show that a large percentage of the freight from Addis Ababa to Djibouti is agricultural products. Although the freight rates charged by the railways are high, the railway is still the main artery of internal and external trade in Ethiopia. The reason for the high freight rates is that the railway line has to pass long distances of uneconomic regions. Moreover there is a lack of local supplies of fuel.[27] In 1980, the Addis Ababa-Djibouti railway carried an average of 450,000 tons of freight. Machinery and equipment and oil products formed the bulk of import items. The main export items were foodstuffs and agricultural raw-materials.

25. Ministry of Planning and Development, Third Five Year Development Plan (Second Draft), *op. cit.*, p. 21.
26. *Ibid.*, p. 23.
27. H.P. Huffnagel, Agriculture in Ethiopia, *op. cit.* p. 88.

The development of a modern highway network in Ethiopia is a recent phenomenon. Due to the vast mountain massif, the ruggedness of the topography and torrential wet season floods, the development and maintenance of an up-to-date road system has remained difficult and costly.[28] Thus most of the transportation of goods has been restricted to pack animals over tracks. The real cost of such transport is very high and much time is wasted on marketing instead of increasing the production. Before 1935 there were only 1,040 kms. of all-weather road tracks which increased to 7,251 kms. in 1967-68, and 16,000 kms. in 1984. Although Ethiopia made immense efforts to improve the road network, the existing system is very small both per thousand of population and per hundred square kms. of area of the country.[29] The existing roads extend mostly to the provincial cities ; these are almost connecting links between rural settlements and main roads. The lack of transport facilities is the main cause of the development of agriculture in belts close to the roads.

Main Agricultural Products

Ethiopia has different kinds of soil, climates and altitude and growing seasons throughout the year. As a result, different kinds of crops are cultivated in the country.[30] It is not possible to locate all the crops cultivated in Ethiopia because of the general topography and the variations of the growing areas. Some of the crops can be grown everywhere while others are confined to specific areas depending on the rainfall and altitude. The small grains are cultivated on higher elevations. Crops like corn and teff are cultivated in areas of medium elevation and those like cotton in somewhat lower areas.

1. *Cereal Crops*

The major cereal crops of the country are teff, sorghum,

---

28. The Technical Agency, IEG, The Resources and Economy of Ethiopia, *op. cit.*, p. 83.
29. Ministry of Planning and Development, Third Five Year Development Plan (Second Draft), *op. cit.*, p. 25.
30. K.A. Cherian, Ethiopia-To-day, *op. cit.*, p. 35,

barley, maize, wheat and millet, *Teff* is the most widely grown grain crop and is the staple food of the majority of the population. It ripens very quickly, needs a little attention, and involves very small threshing losses. The bread made from teff flour is called 'injera'. Its cultivation is most widespread between the altitudes of 1,700 and 2,800 metres. The seed is generally sown at the beginning of the heavy rains in June and it gets ready for harvest in October or November. Two main types of teff are recognised-white and brown-red. The former is suited to the more fertile soil and takes a little longer to mature. Teff, being an important crop accounts for about 30 percent of the cultivated area and about 25 percent of the total production.[31]

Sorghum is more widely grown in the north and east of the country in the lower rainfall areas and used for human consumption and for brewing. It is very popular with the Ethiopian farmers because of its hardiness and easy cultivation. Numerous varieties and types of sorghum are cultivated in the various climatic regions of Ethiopia. It can be grown up to an altitude of 2,500 metres, but the best sorghum producing areas are found around 1,800 metres. Sowing is done in March, April, May, and before the beginning of the main rainy season in July It takes about 7 to 8 months to mature. It is very difficult to store sorghum for a long period since the moisture is easily absorbed again after drying.

Barley is very likely the second most important grain crop in Ethiopia. It is a typical crop in Ethiopia and its zone of cultivation corresponds with that of wheat. Its cultivation is rarely done under 1,800 metres. In south Ethiopia the richest cultivation exists between 1,800 and 2,100 metres, and at 2,700 metres in the Chercher Highlands. In the highest arable plots at 2,900 metres only barley is cultivated. There are many species of barley which have been grown in Ethiopia for many years. It has many uses in the home but the primary use is to make beer (talla). Some barley is used to make injera. It is sown in the beginning of the main rainy season about the end of July or in the beginning of August. Harvest gets ready in November and December.

31. Georgi Galperin, Ethiopia : Population, Resources, Economy, Progress Publishers, Moscow, 1981, p. 180.

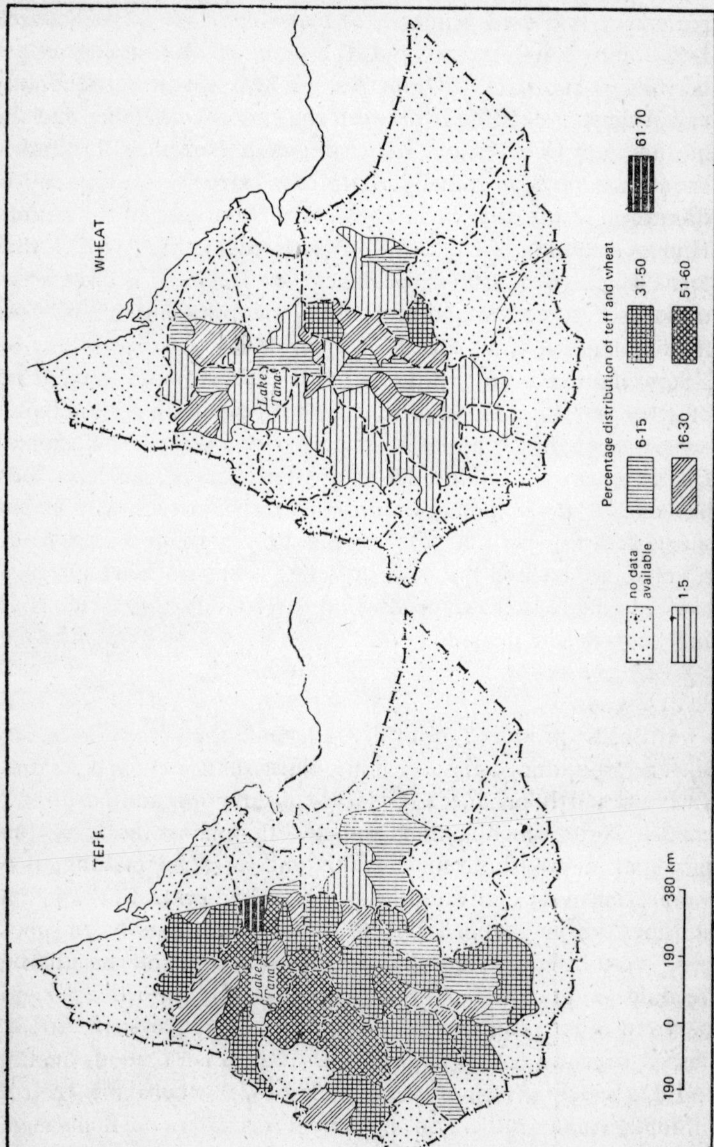

Fig. 2.5 : Crops Distribution (Teff and Wheat)

Wheat is a very important crop in the black soil of the high plateaus. It is used primarily to make flour. It grows between 1,500 and 2,500 metres but the main cultivation centres are between 1,800 and 2,200 metres. Wheat is sown around July and August. Heading is between the end of September and the beginning of October and harvesting is in November-December or in some cases in January. *Maize* is extremely widespread in dry areas, especially in south-western Ethiopia, in the eastern Highlands and in northern Ethiopia. It adapts itself well to that areas which are 2,500 metres above sea level. It is often sown together with cotton and sorghum. The most suitable time for the highlands is from March to May. *Oats* are grown only on a very limited scale at high altitudes. It is found in mixtures of other small grains. It is an excellent grain crop for animals.

Millet or finger millet is sown in many areas in November-December as an additional crop after other cereals have been harvested. It is a simple, drought resistant crop with a very short growing period. During the 1950's attempts were made to introduce rice in the Awash Valley. But no much progress has been made because production is relatively labour-intensive and its domestic demand is low.

2. *Oilseeds*

Oilseeds produced in Ethiopia include such crops as castor, cotton, groundnuts, linseed, noug, rape, sunflower, and sesame. They are partly produced for local consumption and partly for export. Noug which is indigenous to Ethiopia is the country's principal oil crop both for cooking and for soap making. It is more extensively cultivated in northern and central Ethiopia at altitudes varying from 1,800 to 2,500 metres mostly on poor soil. Very little care is given to its cultivation and usually no weeding is practised. It is used extensively in Ethiopia and exported in large quantities. As in many other parts of Africa, linseed is grown for oil in Ethiopia. The oil is of good quality and it is widely grown on the highlands of northern and central Ethiopia usually rather on poor soil or after the fields have been planted for two or three years with grain crops. Little care is given to its cultivation. Sowing is usually done in August and the crop matures in November,

Castor Bean from all indications is indigenous to Ethiopia. The growing area for this crop in Ethiopia is the higher rainfall area and at altitude upto 3,000 metres. In the last two decades, the demand for this crop has greatly increased and this is due to its extensive use as a lubrication oil. The seeds are of very good quality and when processed and pressed yield 40 to 50 per cent oil. Most of the castor seeds at present exported are from wild shrubs in western, central and northern Ethiopia. Other oilseeds cultivated for local use are rape seed produced in the provinces of Arussi, Shoa, Gojam. The plant is cultivated for seeds and leaves, the later being consumed as vegetables. *Sesame* is cultivated in many parts of Ethiopia. It is a high-quality crop containing 45-55 per cent oil. The seeds are partly used as food and are frequently grown together with cotton. Sunflower is grown in high places where there is good rainfall, sunshine and soil. It is cultivated chiefly in Harerge and Arussi. The taste and composition resemble those of safflower which is also cultivated in Ethiopia since ancient times.

Groundnuts are a relatively new crop in Ethiopia and are grown only on a limited scale. The main producing centres are western Eritrea and Harerge. Satisfactory results were reported from the Matahara region. It is grown on altitude which ranges from about 800 to 1,800 metres. Planting is done at the end of June and harvesting takes place in October. Most of the crop is exported and only small quantities are consumed by the people of the country, which have little taste of groundnut. *Cotton seeds* are also produced in large quantities in Ethiopia. The best local varieties are grown on the seacoast. In small farms it is frequently grown with other crops.

Oilseeds in Ethiopia are very important as food, as raw material and as an export item. The total production of all types of oilseeds has been estimated at about 0.8 million tons. The total exports of these seeds account for about 10 per cent of the value of all exports and occupy third place after coffee, hides and skins. It has been estimated that the largest contribution to the diversification of agricultural exports is expected from oilseeds.[32]

---

32. Ministry of Planning and Development, Third Five Year Development Plan (Second Draft), *op. cit.*, p. 47.

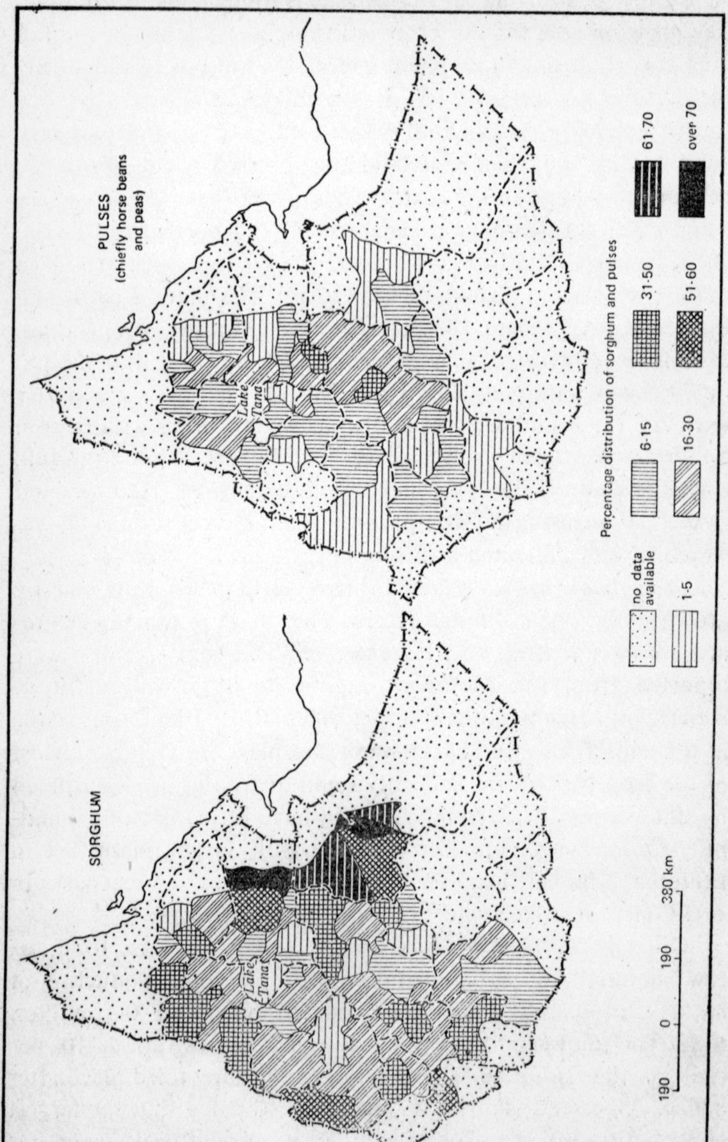

Fig. 2.6 : Crops Distribution (Sorghum and Pulses)

3. *Pulses*

The major Ethiopian pulses are chick peas, field peas, haricot beans, horse beans, and lentils. Though pulses are a daily food in many parts of Ethiopia and many different kinds of pulses are produced for both local consumption and export, complete information is not yet available on this aspect. *Chick Peas* are widely grown in Ethiopia mostly for the local market, and a little for export. The main producing centres are northern and central Ethiopia, but small plots are also found elsewhere. It is usually sown at the end of the rainy season and harvested in or around February. In large areas of the central and northern highlands peas are grown, either in a rotation scheme just after cereals, or on poor, sometimes on badly eroded soil. Peas are locally used for food. Both green and dried peas are also exported.

Haricot bean is mainly a cash crop. There are different types of beans. The small varieties are mainly produced in the Chercher area and the medium ones in Arussi and Shoa. Haricot beans grow best on altitudes ranging between 1,700 and 2,000 metres. Most of them are purchased for processing by canneries. The good qualities are exported. The lentils grown in Ethiopia are small and rather dark coloured. They are mostly locally consumed. A small type of horse bean is widely grown in Ethiopia. Most of it is exported both for human consumption and for processing into cattle food. The sowing and harvesting period for most of the pulses is the same as for chick peas.

4. *Sugar-cane*

Although sugar cane has long been cultivated in Ethiopia, it has been in small areas. Only sugar cane stalks were used for food as confection. In 1954 the Ethiopian Government awarded a concession to the Dutch firm, HVA for the operation of the Wonji Sugar Estate. The altitude of the Wonji plain is about 1,500 metres above the sea level. The climate of Ethiopia in general is not ideal for sugar cane production because of the somewhat cool nights and the very dry seasons. Sugar cane should have warm moist nights and sufficient water. The cool nights cause the canes to be somewhat shorter as compared to the larger sugar-cane areas in other countries.

Rainfall at Wonji is scanty. Sufficient rainfall may be expected in July and August only during which time no supplementary irrigation is required. But during the other months irrigation water must be supplied on a large scale, as the period between October and February is almost completely dry. For this purpose irrigation water is pumped from the river Awash by a pumping station. An advantage of the Wonji climate is that during the prolonged period of drought, the ripening of the cane is satisfactory and the harvesting during November to June is seldom disturbed by rainfall. Intensive cultivation of sugar cane is possible on irrigated soil in the hot areas. But it is so far limited to plantation in Awash Valley.

5. *Cotton*

There are many areas in Ethiopia where soil, climate and altitude are suitable for cotton cultivation. Until recently, Ethiopia used to import cotton for the textile factories in the country. Due to the efforts of the Government cotton plantations have been developed in the lower Awash Valley and in Eritrea. The Tendaho cotton plantation and the plantation in Tessenei in western Eritrea now meet a large part of the demands of cotton textile factories in Ethiopia. As a rule, rainfall decreases while temperature increases. The critical altitude for cotton where the incidence of frost is reduced and sufficient rainfall occurs is roughly at about 1,400 to 1,500 metres. But the main problem at this height is the cold night temperatures in October and November, just after the rains. The most suitable cotton areas are situated below 1,400 metres and above 1,000 metres. Because of the occurrence of malaria, peasant farmers avoid these areas. Hence very little cotton is grown at present here, although they are the most promising and form the bulk of the potential cotton growing areas of Ethiopia. Plantation cotton farming on both small and large scale when accompanied by malaria control measures, will be well suited for these areas. By far the most successful method of growing cotton in Ethiopia at altitudes above 1,000 metres on a plantation basis is the use of the supplementary irrigation. By this means it is possible to plant in March or April before the rains begin, thereby obtaining the crop in six-and-a-half to seven months. In this way the best use is made of the available rainfall.

6. *Coffee*

Coffee is the most important crop of Ethiopia as a provider of foreign exchange. Ethiopia is the original home of coffee, the name having been derived from Kefa, a province where coffee has been a wild crop. This crop is produced in Ethiopia under various conditions.[33] A large part of the crop is still harvested from the so-called coffee forests. They are found chiefly in the provinces of Kefa, Illubabor, Wollege and Gemu-Gofa.

In the provinces of Kefa, Sidama and Gojam, established plantations of sizeable acreage are most noticeable. They are mostly grown under shade and without any systematic planting, pruning and upkeep. Most of these plantations are found in the provinces of Arussi and Sidama and Chercher district of the province of Harerge. Some plantations in Harerge province are unshaded and often irrigated and given very great care.

Tests of Ethiopia coffee of different types have shown that coffees is of high quality when properly harvested and processed. It can even compete with the best coffee producing countries in other parts of the world. It is mostly used abroad for blending. The abundant and regular rainfall in south-west Ethiopia which comprises provinces of Kefa, Ilubabor and Wollega is favourable for the growth of high grade coffee. These areas are found at elevations which range from 1,600 to 2,100 metres—the best growing areas in Ethiopia due to the combination of rainfall and temperature. At these highland levels, the temperature is rather uniform throughout the year and there are little variations. At altitudes below 1,600 metres with regular rainfall, the climate becomes hot. It does not mean that no coffee trees are found on lower altitudes. There are some areas for example Mizan Tafari between 1,200 and 1,300 metres where superb trees with abundant fruits have been found but at the same time some diseases are also found in these areas.

---

33. Ministry of Information, Agriculture in Ethiopia, Book II, *op. cit.*, p. 11.

At present, it is estimated that coffee production amounts to about 180,000 tons.[34] Ethiopia is also a member of the International Coffee Organisation which limits the exports by quota system At the moment this annual quota amounts to about 74,000 tons which is less than its exportable capacity. The quality of coffee, which is the Arbica type helps determine its demand abroad. To improve the quality of Ethiopian coffee some producers are now using the wet method for curing coffee beans. With this wet method the coffee gets a better price than the sun dried. Such methods are very essential to capture world markets.

The Ethiopian Government has endeavoured to improve and extend the cultivation of coffee. In 1952, the Imperial Government issued a proclamation providing for the licensing of coffee cleaners and graders by which all coffee for export must be inspected by the National Coffee Board. The moisture content must not exceed 11.5 per cent. In 1957 the National Coffee Board was given much powers to deal with the problems of coffee cultivation, processing and export. In 1959 regulations were issued requiring that all licensed coffee cleaners and graders should apply a specified classification system. In May 1979, a special Ministry of Coffee and Tea Development was established. The marketing section, called the Ethiopian Coffee Corporation was also established to facilitate the activity of coffee trade. Thus, considerable efforts are being made for improving the quality of coffee since it provides a large amount of the country's foreign exchange earnings.

7. *Tobacco*

Tobacco has been known in Ethiopia for several centuries. Turkish tobacco is thought to have reached the country in the16th century. American brands are now also cultivated. Till recently, tobacco was grown exclusively by the Ethiopian Muslims [35] There are no special tobacco plantations and it is grown with other crops. The principal cultivation areas are Northern

34. Georgi Galperin, Ethiopia-Population, Resources, Economy. *op. cit.*, p. 189.
35. *Ibid.*,

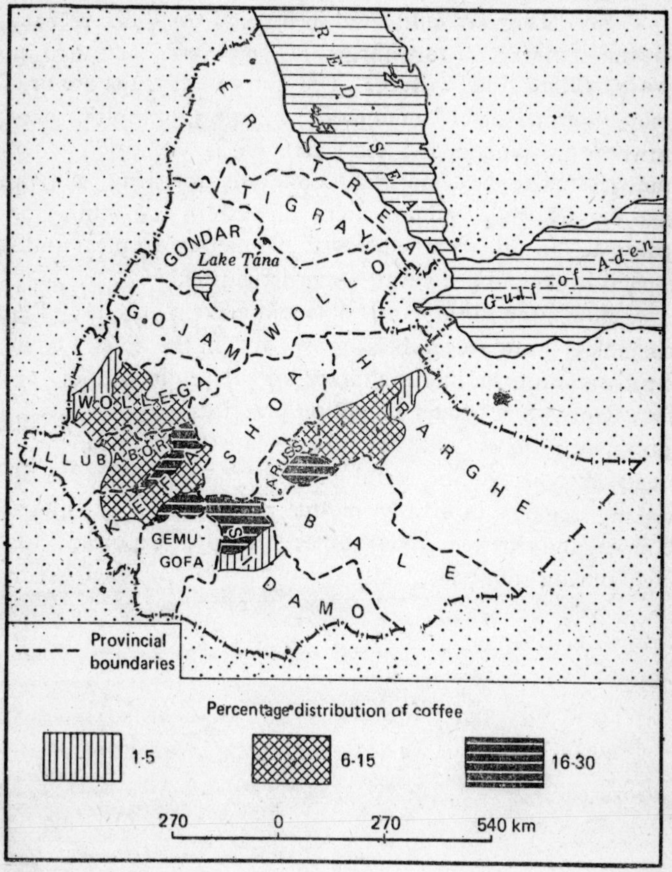

Fig. 2.7 : Principal Coffee Growing Areas

Sidamo, Southern Shoa, Eritrea and Harerge. Cash-crop production is concentrated in Shoa and Sidamo areas. A large tobacco plantation is presently being cultivated in the valley of River Billate. Two cigarette-manufacturing factories use local and imported blends. In future, tobacco may also become an export item.

8 *Fruits and Vegetables*

A great variety of fruits are cultivated in Ethiopia though

they are seldom included in the regular diet of the Ethiopian population. They are cultivated in big and small farms throughout the country. Fruit farms of large size are found in the Awash Valley and Eritrea. The important fruits are orange, lemon, grape fruit, mangoes, bananas, apples, papaya, strawberries, pineapples, etc. Banana is exported to other countries. The best natural conditions for fruit cultivation exist in the high rainfall areas of western and south-western Ethiopia. In other parts below 2,200 metres, tropical and sub-tropical fruits can grow well under irrigation.

Although the Ethiopian people still make a little use of green vegetables, most vegetables grow well in the upland country. The dry season is a limiting factor for the production of vegetables. Only when proper irrigation practices are introduced, the Ethiopian farmers will be able to satisfy demands which are already increasing. Successful gardens for vegetables are found in many regions especially in the provinces of Shoa, Arussi, Harerge and Eritrea. Vegetables are also exported to the neighbouring countries.

PART II

AGRICULTURE BEFORE THE REVOLUTION

CHAPTER THREE

Growth in Agricultural Output and Area

IT IS AGAINST this background of natural and economic conditions prevailing in Ethiopia that an attempt has been made to assess the performance of the agricultural sector during the three Plan periods i.e., from 1957 to 1974. Since agricultural production depends upon the area under cultivation and yield per unit of land, these two variants have also been analysed.

Until recently, very little was factually known about the area under crops, yield per unit of land and total production of each crop. Estimates had been made on a national basis over a period of years based on very meagre sample survey data.[1] However, at the time of formulation of the Second Five Year Development Plan, some systematic efforts were made to calculate areas, yield and production of different crops. But

1. Dr. Willis G. Eichberger, Food Production and Utilisation in Ethiopia, 1958 EC, Addis Ababa, May, 1968, p. 1.

all the figures concerning these three aspects were estimated. The basis of production estimates were the estimations of per capita consumption of various groups of agricultural products. The data on per capita consumption of agricultural products were obtained through various partial surveys on consumption and nutrition. The result was that very crude estimates could be made, especially of individual crops during the Second Plan.[2] The Third Plan also used this methodology with some corrections. Most of the estimates were revised after data from different CSO-Rural Surveys were made available. The CSO was able to conduct surveys in 12 of the 14 provinces. Moreover, the Ministry of Agriculture had also followed the 'cutting out crops' procedure to determine the actual yield per hectare. The main limitation of this analysis is that the statistical data relating to production, yield and area under crops for different years under three Plan periods are not available at one place. Different sources have to be consulted for this purpose, with the result that large variations have been found at the time of analysing the yearwise date. However, every effort has been made to minimise such variations.

Rise in Output

Before the implementation of the First Five Year Plan (1957 to 1961) the rate of growth of major agricultural commodities was estimated at about 1.6 per cent per annum as against 1.5 per cent annual increase in population. It was assumed at that time that the increase in agricultural production was slightly above that of the total increase in population.[3] The rate of growth of cereal crops was estimated at about 1.2 per cent. The maximum growth was observed in the case of fruits and vegetables (2.1) followed by pulses (1.6).

Realising that agricultural development could provide better and more varied diet for the population, more raw-material for domestic manufacturing industries and larger

2. Ministry of Planning and Development, Third Five Year Development Plan (Second Draft), *op. cit.*, p. 1, Appendix.
3. Imperial Ethiopian Govt., Five Year Development Plan, 1957-61, *op. cit.*, p. 10.

surpluses for export, two important objectives were fixed in the First Plan. Firstly, to correct the basis for the more dynamic long term development of agriculture and secondly to take measures that would achieve during the planned period a sizeable increase in agricultural production. So the target for the growth of agricultural production was fixed at about 2.2 per cent per annum.[4] The break-up in Table III.1 shows that the production of cereals was expected to grow at the rate of

TABLE III.1

Production of Major Agricultural Commodities from 1954 to 1961

(Thousand Tons)

Crop	1954	1957	1961	1957 over 1954	1961 over 1957
Cereals	4323	4526	5031	4.7	11.2
Teff	2546	2664	2831	4.6	6.3
Wheat	191	201	310	5.2	54.2
Others	1586	1661	1870	4.7	12.6
Oilseeds	251	264	285	5.2	8.0
Noug	186	195	305	4.8	5.1
Others	65	69	80	6.1	16.0
Pulses	487	518	555	6.4	7.1
Horse Beans	110	115	124	4.5	7.8
Chick Peas	169	178	191	5.3	7.3
Lentils	90	101	115	12.2	13.9
Others	188	124	125	5.1	0.8
Coffee	141	142	156	0.4	10.4
Fruits & Vegetables	180	195	239	8.3	22.6

Source : Imperial Ethiopian Government—Five Year Development Plan, 1957-61, Addis Ababa, pp. 165-66.

4. The production of all the agricultural commodities has been added for the years 1957 and 1961 in order to derive percentage changes in 1961 over 1957 for the agricultural sector as a whole.

2.2 per cent per annum during the planned period. The production of wheat was expected to rise at the rate of 10 per cent per annum. It was also decided to take measures to increase the output of raw material and surplus marketing crops to meet the large requirements. The production of oilseeds and pulses was expected to increase at the rate of 1.6 and 1.4 per cent respectively per annum. Since coffee has for a long time been cultivated in many parts of the country and still it is the only plant grown more extensively for the market and its exports have been constantly increasing, it was expected that coffee would expand at the annual rate of 2.1 per cent. The production of fruits and vegetables was expected to increase at the annual rate of 4.5 per cent.

Statistical data showing trends in the production of different crops year-wise is not available in detail and it is difficult to arrive at the conclusions about the actual progress made during the First Plan period. However, it is claimed that the targets concerning agricultural production had been fulfilled.[5] According to very rough estimates, the production of cereals had increased by 2 per cent during the Plan period. The production of other crops had also registered some increase. But this increase in the production of cereals was not enough to meet the growing demand of the population and consequently Ethiopia became an importer of wheat (45,000 tons in 1961) instead of being an exporter as it was 5 to 6 years ago.[6] The growth of industrial crops which was anticipated by the Plan was also not fully realised. A large import of cotton had to be made to meet the growing consumption of new textile factories. In the case of other crops, the growth of output was such as to meet the increasing demand arising from the growth of population and some modest requirements of the industry.

Output during the Second Plan

During the Second Five Year Development Plan (1963-67) it was realised that since the economic development would be

5. Imperial Ethiopian Government, Second Five Year Development Plan, 1963-67, *op. cit.*, p. 44.
6. *Ibid.*, p. 45.

faster in the course of the Plan, "it is obvious that agriculture has to speed up its pace of growth, so as to be able to satisfy the growing demand of agricultural commodities."[7] On the basis of the estimated population growth and the rise in per capita consumption in various foodstuffs, it was estimated that the total food consumption in 1967 would be about 11 per cent higher than it was in 1962. As such it was planned to increase the production of cereals in relation to the population by 530,000 tons during the Second Five Year Plan to satisfy the domestic need which would arise during the coming years. In order to meet the increased demand and at the same time to decrease imported food products as well as for the planned increase of exports of agricultural products, agricultural production was to increase at the rate of 2.3 per cent as indicated in Table III.2.

TABLE III.2

Production of Major Agricultural Commodities in 1967

Item	Total Increase (000 tons)	Index (1962=100)
Cereals	530	111
Oilseeds	73	120
Pulses	51	110
Vegetables	76	115
Fruits	17	130

Source : Imperial Ethiopian Government, Second Five Year Development Plan, 1955-59 EC (1963-67), Addis Ababa, 1962, p. 111.

It was also estimated that the production of oilseeds would grow at the rate of 3 per cent per annum and the total increase in the production would be 73,000 tons. This increase was required to improve nutrition, provide raw material for domes-

7. *Ibid.*, p. 46.

tic industries and increase exports. The increase in the production of pulses was expected to be about 2 per cent per annum in order to fulfil the needs of domestic consumption and to increase export and also for the new canning industry which would develop in the course of the planned period. The increase in the production of vegetables and fruits was expected at the rate of 3 per cent and 6 per cent respectively during the same period.

Table III.3 briefly shows some of the major targets of the Plan and their achievements in terms of increased production over the five years. Although there was an overall increase in the production of all the commodities, targets were not fully realised in most of the cases. The cereal production increased at the rate of 2 per cent per annum. The production of oilseeds and cotton reached levels not much more than half of the targets set out in the Plan. It was only in the case of coffee and pulses that the Plan targets could be surpassed. The overall rate of growth for agricultural production was 2.1 per cent per

TABLE III.3

Second Five Year Development Plan-Targets and Achievements

Products	Targets (000 Tons)	Percentage Implementation
Cereal Production	530	91
Oilseeds	98	53
Pulses	51	105
Vegetables	76	96
Fruits	17	88
Sugarcane	250	82
Coffee	18	176
Cotton-lint	12	50
Coarse fibre	9	29

Source : Imperial Ethiopian Government, Third Five Year Development Plan, 1961-65 EC, (1968-1973), p. 186.

annum.[8] But on the other hand some of the areas of the country experienced crop failures repeatedly during this period and consequently hardship to the population. In 1963 and 1964 the production of agricultural commodities increased normally and there was not much shortage of food commodities. But the main economic characteristic of 1965 was the shortage of the agricultural commodities such as teff, maize, millet, etc., which were domestically produced and consumed. The country had to import wheat, wheat-flour and durrah to make up the deficit. These imports consisted of about 9,000 tons purchased and about 17,000 tons received as aid during 1965.[9] In general, agricultural production in 1966 was not reported to have been worse than the previous year. The increase in output in certain areas was of course offset by drought in other regions. Import of grain and flour amounted to be about 19,000 tons of which 10,000 tons of wheat grains had been obtained from the USA against a credit of Eth. $ 1,095,000.[10] The production of agricultural commodities continued to expand in 1967 as in the previous year at a moderate rate. But towards the end of the year, again the occurrence of the unusual weather conditions[11] affected the country and production in turn.

Output during the Third Plan

Realising that agricultural progress was slow during the first two Plan periods, emphasis was once again put on the 'same foundation-laying and preparatory measures' during the Third Plan (1968-73). To a larger extent the Third Plan period was regarded as a preparatory period designed to ensure more rapid and sustained agricultural development in the subsequent plans. "The success of the Plan, and of Ethiopia's sustained development, depends above all on the achievement of steady progress in agricultural output and productive efficiency. Ethiopia needs to produce steadily the expanding supplies of foodstuffs to meet

8. Imperial Ethiopian Government, Third Five Year Development Plan, 1968-73, *op. cit.*, p. 185.
9. National Bank of Ethiopia, Annual Report, 1965, Addis Ababa, p. 6.
10. National Bank of Ethiopia, Annual Report, 1966, Addis Ababa, p. 6.
11. National Bank of Ethiopia, Annual Report, 1967, Addis Ababa, p. 6.

the growing money demand for food from its rapidly increasing urban population. The present nutritional levels of much of the population must be improved. Ethiopia also needs to expand agricultural exports in order to pay for the growing volume of imports demanded by a growing economy. Again, as industrialisation progresses, so will the demand increase for essential raw materials, many of which imported. Conversely, an agricultural sector which is not growing fast can only perpetuate the low living standards of the majority of the population because the low productivity and income of the farmers provide neither an attractive market nor allow any significant savings to be made."[12]

TABLE III.4

Projected Gross Output of Selected Agricultural Products from 1960 to 1965 EC (1967-73)

(in Thousand Tons)

Products	1967	1973	Percentage increase in 1973 over 1967	Rate of Growth
1. Cereals	5486.8	6226.2	13.5	2.6
—Teff	1304.3	1402.3	7.5	1.5
—Wheat	760.0	922.5	21.4	4.0
—Others	3422.5	3901.4	14.0	2.7
2. Oilseeds	437.4	607.9	38.6	6.8
3. Pulses	599.6	713.6	19.0	3.6
4 Coffee	160.0	185.0	15.6	2.9
5. Sugarcane	858.0	1375.0	60.3	9.9
6. Cotton fibre	10.4	26.0	150.0	20.1

Source : Imperial Ethiopian Government, Third Five Year Development Plan 1961-1965 EC (1968-73), Addis Ababa, p. 203.

12. Imperial Ethiopian Government, Third Five Year Development Plan, 1968-73, *op. cit.*, p. 181.

The total output of agriculture was expected to grow at an annual rate of 3.1 per cent. The rate of growth of cereal crops had been fixed at about 2.6 per cent per annum. The maximum rate of growth was to be achieved in the case of wheat i.e. 4.5 per cent per annum. It was expected that the output of oilseeds would grow much faster than agricultural output as a whole—about 6.8 per cent per annum. The export of oilseeds was to increase by three times during the Third Plan period. Pulses, as an important food and export items, were to grow at an annual rate of 3.6 per cent. Industrial crops like sugarcane and fibre were expected to grow at an annual rate of 9.9 per cent and 20.1 per cent respectively. The growth rate of coffee was fixed at 2.9 per cent per annum.

The Third Plan period which was extended upto 1974 can be divided into two parts. During the first part of the Plan period, i.e., 1969 to 1972, as shown in Table III.5, the economy registered some positive rate of growth both in cereal and non-cereal crops. The production of cereals increased by 4.1 per cent during this period. The output of wheat went up by 8.4 per cent followed by maize by 7.0 per cent. The position regarding non-cereal crops was also satisfactory. The production of peas and sesame seeds increased by 9.3 per cent and 52 per cent respectively. The situation took a serious turn only in the second part of the Plan period i.e. 1972 to 1974. During this period, the production of cereals decreased by 7.4 per cent. The maximum decrease was in the case of wheat. Although the production of non-cereal crops increased during this period, the rate of growth was comparatively less. During the entire Plan period, the production of cereal crops went down by 3.6 per cent. As a result of this, a severe famine swept into the northern provinces of Wollo and Tigre and thousands of people died. According to one estimate about one-fifth of the population of Wollo province had died in 1973. The number of deaths from starvation in Ethopia might reach 500,000.[13] The fali in agricultural production in those areas was not simply due to lack of rainfall only. Even

13. Jack Sheperd, The Politics of Starvation, New York, 1975, p. 37.

in those parts of the country not directly affected by drought, the picture was far from encouraging. At the end of June, 1973 retail price index (1963-100) in Addis Ababa had risen to 150.2 compared to 134.5 a year before. Retail prices for cereals, fruits and vegetables and pulses almost doubled from what they were in August 1972.[14] Even after the declaration of National Emergency in November 1973 by the Government, people continued to starve in Ethiopia.

TABLE III.5

Production of Important Agricultural Crops During Third Plan Period

Crops	1969	1972	1974	1972 Over 1969	1974 Over 1972	1974 Over 1969
	(000 Tons)					
Cereals	5752	5989	5545	4.1	—7.4	−3.6
—Wheat	808	876	750	8.4	—14.4	—7.3
—Barley	1496	1601	1400	7.0	—12.5	—6.4
—Maize	909	971	900	6.8	— 7.3	—1.0
—Other Cereals	2539	2641	2495	4.0	— 5.7	—1.9
Check Peas	181	198	198	9.3	—	9.3
Dry Broad Beans	139	148	154	6.5	4.0	10.8
Coffee	170	175	180	2.9	2.9	5.8
Lentils	106	110	113	3.8	2.7	6.6
Sesame Seeds	69	105	110	52.1	4.7	59.4
Sugarcane	NA	1330	1450	NA	9.0	9.0

Source : FAO ; Production Yearbook, 1971 Vol. 25. Rome, 1972 and Production Yearbook, 1974, Vol. 26 No. I, Rome, 1975.

Though it is not very sensible to derive average annual rate of growth of agriculture for the entire three Plan periods, the foregoing analysis reveals that agriculture in Ethiopia had suffered from chronic stagnation during each of three plan

14. The Economist Intelligence Unit, Quarterly Economic Review of Ethiopia, Uganda and Somalia, Quarter IV, 1973, *op. cis.*, p. 21.

periods.[15] Since agriculture accounts for 60 per cent of the GDP, the stagnation of the agricultural sector had been the major constraint on economic growth. Further, the annual rate of growth of cereal crops was very low as compared to non-cereal crops. Consequently, the shortage of food did emerge in 1973.

Predominance of Area over Yield

The increase in agricultural production in due to either an increase of area under cultivation or an increase in yield per unit of land or a combination of both. During the First Plan period, both the methods were considered for the expansion of agricultural output. The Plan stated that "the main obstacle to extend the cultivated area is the low productivity of labour. To solve this question there is no need to bring about fundamental changes in the present method of production nor it is desirable to give up present inefficient tools... Most of the proposed tools can be produced in the country and the people who use them would not require complicated instructions......There is no doubt that in due course modern machines such as tractors, drillploughs and threshing machines will be introduced on the larger farms and plantations, but for the time being the introduction of the simple tools already referred to would yield better results as they can be introduced on the farms of a large number of agricultural producers."[16] Although the importance of improved yield was recognised for the overall expansion of agricultural production, it was assumed that the increase in yield per unit of land would not have significant impact on the production during the First Plan period. In almost all the commodities, it was the increase in area that contributed the most to the increase in production.

During the Second Plan, emphasis was again laid to achieve the targets of agricultural production by increasing the area

15. Zoltan Gyenge, Ethiopia on the Road of Non-Capitalist Development, Studies on Developing Countries, Institute for World Economics of the Hungarian Academy of Science, Budapest, 1976, p. 22.
16. Imperial Ethiopian Government, Second Five Year Development Plan, *op. cit.*, p. 20.

under cultivation and improving the yield per hectare. "Considering the low yields which result from extremely inefficient farming methods and tools, it is also possible to raise the yields by area units considerably on a dry farming system. There are large areas of land that can be irrigated to allow high and stable yields and at least two harvests yearly. There are still large waste cultivable lands, particularly in the sparsely populated regions, which could be cultivated."[17] Table III.6 shows the expected rate of growth of production and area under important crops during the Second Plan period. Although the Second Plan considered the significance of improved yield per hectare alongwith the increase in area under cultivation to achieve the agricultural targets, it would be clear from the table that the increase in area was expected to contribute most to the increase in production. In case of cereals, the increase in area

TABLE III.6

Growth Rate of Production and Area Under Important Crops During Second and Third Plan Periods

Crop	Rate of Growth Per Annum			
	Second Plan		Third Plan	
	Production	Area	Production	Area
Cereals	2.09	0.62	2.60	1.30
—Teff	1.24	0.62	1.50	0.90
—Wheat	4.15	3.20	4.50	1.30
Oilseeds	3.05	2.82	6.80	5.00
Pulses	1.92	1.38	3.60	1.30
Sugarcane	5.70	3.40	9.90	9.00
Coffee	3.60	0.80	2.90	0.60
Tobacco	7.75	7.50	NA	NA
Vegetables	2.00	1.61	NA	NA

Source : Imperial Ethiopian Government, Second Five Year Development Plan (1962-67) Statistical Tables for 1962-67, and Third Five Year Development Plan (1968-73) Statistical Appendix Part I & II, Addis Ababa, 1969.

17. Imperial Ethiopian Government, Second Five Year Development Plan, *op. cit.*, pp. 109-110.

was expected to contribute only 0.62 per cent to realise 2.09 per cent growth. The remaining estimated increase would be the share of increased yield. The increase in production of coffee was to come from the increased yield. But in the case of other non-cereals, area was called for to contribute the most to achieve the desired production goals. The entire increase in the production of tobacco was to be realised through the increase in area under cultivation.

The real importance to the improved yield per unit was given in the Third Plan. "Approximately 13 million hectares or 11 per cent of the nation's land area, are properly defined as cultivated land. Although not true of the potential settlement areas, it is a fact that a significant extension of much of this area is becoming more difficult and increasingly unrewarding in social and economic results."[18] It was therefore imperative to raise the productivity of land to have an overall increase in agricultural production. The Third Plan proposed three important measures. First, the policy of the Plan was to push ahead as rapidly as possible with research and experimentation to gain technical improvement and the result of this work must be fed to the farmers rapidly in such a way that they could use it beneficially. Among the main lines of action would be seed-bed preparation, storage methods, selection of better breeds and control of pests and diseases. It was also planned to introduce more efficient implements and tools and establish 10 multi-purpose agricultural stations for demonstration purposes. Secondly, the promotion of the use of fertilisers was also an important aim of the Plan. The FAO's 'Freedom From Hunger' Campaign had enabled several hundred trials to be conducted in some of the areas. Thirdly, the most promising approach to the successful development of strategically selected areas was the "package programme." This simply means "the policy of concentrating development efforts in given areas of not too great size utilising all the materials and human resources required to change traditional

18. Imperial Ethiopian Government, Third Five Year Development Plan, *op. cit.*, p. 183,

and inefficient methods."[19] It was assumed at that time that all these measures would increase the yield per unit of land and this increased yield along with the increase in area would enable the Ethiopian economy to raise agricultural output to a satisfactory level.

Table III.6 also shows the projected growth of production and area under important crops during the Third Plan period. The annual rate of growth of area under cereal was expected to be 1.3 per cent to achieve 2.6 per cent growth in their production and remaining had been estimated as the contribution of increased yield. In the case of wheat the position was entirely different. The annual rate of growth of area under cultivation was estimated to be only 1.3 per cent to have a growth of 4.5 per cent and the contribution of yield was to be maximum. Similarly the major part of the increase in the production of coffee was to come from the improved yield. The contribution of area was estimated to be only 0.6 per cent for the achievement of growth rate of 2.9 per cent. The contribution of area was to be maximum in the case of sugarcane where the entire increase in production was to come from the increase in area under cultivation.

As far as the actual contribution of area under cultivation and the yield per unit of land to the total agricultural production is concerned, the position was quite different. As mentioned earlier, during the First Plan period, the entire increase in production of agricultural commodities was to come from the increase in area under cultivation and the contribution of yield was negligible. But during the Second Plan period, the yield played a dominant role in the production of almost all the commodities. During the period from 1961-65 to 1968 which covers the period of the Second Plan, there is some positive correlation (both contributed to the production) between area and yield, although the contribution of yield was less than the area. In the case of cereals in general and wheat in particular, it was assumed that during the Second Plan the yield would contribute maximum to the production. But actually as shown in Table III.7 the area under cereal crops increased

19. *Ibid.*, pp. 193-194.

TABLE III.7

Actual Increase in Area and Yield during the Second and Third Plan Periods

Crop	Area					Yield				
	1961-65	1968	1974	1968 over 1961-65	1974 over 1968	1961-65	1968	1974	1968 over 1961-65	1974 over 1968
Cereals	6747	7208	7565	6.8	4.9	728	767	733	5.3	−4.4
—Wheat	928	1029	1100	10.9	6.4	714	739	682	3.5	−7.7
—Barley	1612	1714	1750	6.3	2.1	820	853	800	4.0	−6.2
—Maize	777	838	900	7.8	7.4	956	1051	1000	9.9	−4.8
Sugarcane	5	6	9	20.2	50.0	129008	147778	161111	14.5	9.0
Pulses	850	N.A	999	N.A	17.5	701	N.A	742	N.A	5.7
Coffee	561	N.A	625	N.A.	11.4	253	N.A.	288	N.A.	13.8

Area in 1000 ha., yield kg/ha.

Source : FAO/UN, Production Yearbook, 1972, Vol. 26, Rome, 1973 and Production Yearbook, 1974, Vol. 28, No. I, Rome, 1975.

by 6.8 per cent whereas the contribution of yield was only 5.3 per cent. In the case of wheat, the area increased by 10.9 per cent as against only 3.5 per cent increase in yield. The maximum increase in area was under sugarcane, i.e., 20 per cent and the contribution of yield was 14.5 per cent.

During the Third Plan period, the correlation between yield and area was negative in the case of cereals. The area under cereal crops increased by 4.9 per cent during the period from 1968 to 1974 whereas the yield decreased by 4.4 per cent. In case of wheat, barley and maize the area under cultivation increased while the yield decreased as indicated in Table III.7 It was only in the case of non-cereal crops where the correlation was positive and both yield and area contributed to the production of such crops. In the production of sugarcane the contribution of area and yield was 50 per cent and 9 per cent respectively. But overall contribution of yield for all the crops was much less than the area under cultivation both during the Second and Third Plan periods.

Commercial Sector

The commercial sector had been assigned a special role to play in the development of agriculture during the pre-revolutionary period. In fact the growth rate of agriculture was calculated on the expected expansion of the peasant subsistence sector and peasant monetary sector on the one hand and the commercial sector on the other hand. The development of commercial farms was required for two important reasons : First, the great majority of Ethiopian peasants produce mostly for themselves and therefore offer small surpluses for sale in the market. Commercial sector would produce large quantities for export purposes. Secondly, it would not be possible to realise production targets from the peasant sector exclusively. Important production targets would have to be achieved from the commercial farms based on modern technology and efficient equipment.

During the First Plan period, the commercial farms of course did not have a decisive influence on the growth of production because they were at the initial stages of their development and could not be utilised to their full capacity.

The systematic development of these farms took place during the Second Plan.

The targets of agricultural output from the peasant sector had been more or less fulfilled. Although the Plan targets for agricultural production from commercial farms could not be fully met, these farms had contributed substantially to the production of raw materials such as cotton and sugarcane for domestic industries and fruits and vegetables for export. Most of the commercial farms had been engaged in coffee and oilseed production and these farms had decisive effect on the total production. For the Third Plan, the development of commercial farms was the only way to increase in the relatively short time the production of important raw material for domestic industries and some commodities for export.[20] Therefore, heavy reliance was placed on these farms for the achievement of targets for agricultural output. Table III.8 shows that the rate of growth of 3.1 per cent in agricultural output during the Third Plan period was calculated on the expected expansion of the peasant subsistence sector by 1.8 per cent and peasant monetary sector by 3.8 per cent and by commercial farms by 12.6 per cent. It was also estimated that the increase in the production of peasant sector would be slightly over the projected growth of population.

Detailed information about the progress achieved by these commercial farms is not available for the Third Plan period. But it is a fact that these farms were able to produce sufficient agricultural goods for export purposes. For example, exports of oilseeds and pulses increased tremendously during the famine period. When Ethiopia was unable to produce sufficient food to meet the domestic needs of its population, it had recorded increase in annual production of agricultural goods for export.[21] These farms turned fertile lands into the production of goods for external markets. In Awash Valley, they had brought about 25 per cent of the total irrigable area under

20. Imperial Ethiopian Government, Third Five Year Development Plan, *op. cit.*, p. 191.
21. Peter Koehn, "Ethiopia : Famine, Food Production and Changes in the Legal Order," African Studies Review, Mass. Vol. XXII, No. 1, April, 1979, p. 52.

TABLE III.8

Projected Value of Agricultural Production During Second and Third Plan Periods

(In million Eth. Dollars)

Sector	Second Plan					Third Plan		
	1962	1967	Annual Rate of Growth (Target)	Actual Production in 1967	Actual Rate of Growth	1967 *	1973 *	Rate of Growth
Total Agriculture :	1743	1955	2.34	1946	2.34	2015	2347	3.1
(a) Peasant Subsistence Sector	1324	1438	1.68	1439	1.68	1462	1596	1.8
(b) Peasant Monetary Sector	336	396	3.35	393	3.35	413	497	3.8
(c) Commercial Farms	84	121	7.60	114	7.00	140	254	12.6

*Revised Figures.

Source : Imperial Ethiopian Government, Second Five Year Development Plan, Statistical Tables for 1962-67 and Third Five Year Development Plan, 1961-65 EC (1968-73), Addis Ababa, p. 202.

cultivation. About 60 per cent of the cultivated area in the Valley was devoted to cotton cultivation and another 22 per cent to sugar plantation.[22] This was made possible by the government through various concessions. To encourage commercial farming, the government exempted tractors, pesticides and fertilisers from import duties. Further, large agrarian investments were given income tax relief for three to five years. The government also showed its willingness to construct new roads linking Addis Ababa with important commercial centres.

But in contrast, the government demonstrated much less interest in the progress of peasant sector. In fact, government policies hampered the peasant agricultural development and facilitated its exploitation. It never thought seriously to remove the defects of institutional factors. "Although 90 per cent of the population were 'active' in agriculture, the country was faced with a food shortage. The techno-economic development of production was hampered by the conditions of feudal exploitation. The 'development plans' and decrees of the Government served the expansion of the large farms."[23]

22. *Ibid.*, p. 53.
23. Zoltan Gyenge, Ethiopia on the Road of Non-Capitalist Development, *op. cit.*, p. 22.

CHAPTER FOUR

Land Tenure System

DURING the emperor's regime, the agrarian structure of the Ethiopian economy had indeed the features which were serious obstacles to agricultural progress. The out-dated system of land ownership and tenure was one of them. It was the sum total arrangements whereby land was owned, rented, or otherwise held for use. The system was not only defective but also in no way inducive to the growth of agricultural sector.

The land and in fact the revenue in Ethiopia were in the hands of four category of persons, "firstly, the sovereign who owned crown lands and who, one way or the other, dispensed land to the other three categories ; secondly, the royal family and the nobility in general, who could receive land either in perpetuity or temporarily or be given the proceeds of taxation from gult lands ; thirdly, the church, or more correctly individual religious establishments or ecclesiastics who might possess gult, in effect the taxes from certain land or actual land ownerships ; fourthly, the peasants who could hold land permanently as rist, whether obtained from the sovereign or inherited on the basis of traditional ownership or on a temporary basis in return for specific services rendered to the sovereign or to another landlord."[1]

There was no clear-cut demarcation between the land of the state and that of sovereign and lands of the state and the imperial were jointly called as "crown land." This was done to conceal the extent of imperial lands and the higher amount of revenue collected there from. The best lands in most of the provinces were the part of the imperial lands. The emperor distributed land among the various classes in the society. The recipient of the land paid tax,...tithe...and other required services to the emperor. All other lands which were not covered by right of private ownership were government lands. The government granted such lands to individuals as hereditary taxed land and as untaxed land to co-operatives, foreigners and various settlement and developmental agencies. In 1974 the crown lands constituted about 42 per cent of the territory's land and were mainly in the southern part of the country. But unfortunately, a large part of the crown lands was either under-utilised or unutilised before the Revolution.

The title to land held by the nobility derived from the emperor. It was one of the traditional imperial rights that the emperor could grant to his subjects the land. Such lands were given to those members of the nobility who had distinguished themselves in the service of the state. They were given sufficient land to use as long as they remained in office. But with the weakening of the authority, land became hereditary and the nobility became owners of the land and treated the tenants as feudal subjects. Such feudal relations formed the foundation of the socio-economic structure of the economy and "permeated virtually all types of landownership and land tenure...The feudal stood between the peasants and the State as taxman and middleman during sales of farm products, between the peasants and the church as head of the parish, and between the peasants and the Law as the judge."[2]

1. Richard Pankhurst, State and Law in Ethiopian History, Published by Haile Selassie I University, Addis Ababa, 1966, p. 31.
2. Georgi Galperin, Ethiopia : Population Resourees Economy *op. cit.*, p. 146.

The feudal lord was also the owner of a considerable part of the produce of the tenant which was as high as 75 per cent. Further, landlords as middlemen paid only 35 per cent of the selling price to the tenants. The introduction of contracting system from the mid 1960s led to the further worsening of the position of the tenants.[3] Under this system, rent was fixed beforehand without taking into consideration weather conditions. The landlords had full right to evict the tenant if he was unable to cultivate the whole plot. Most of the tenure arrangements were verbal and the tenants were supposed to render various services to the landlords. In 1974, about one-third of the country's land belonged to this group.

Ever since the introduction of Christianity, the Ethiopian emperors had generously granted large areas of land to individual churches and monasteries. They used to bestow upon the church one-third of the lands they had annexed through conquests. Such lands were on permanent basis and churches were considered for all purposes as the owners of their holdings. They leased their land on temporary or permanent tenancy to the peasants. The tenants paid tax and tribute and rendered service obligations to the churches. Thus the church's relation with the tenants was essential a feudal one. In 1974 the church land was about 17 per cent of the country's land. But about 50 per cent of the land was not under cultivation.

The peasant also had a right to possess the land tilled by him. His title to the land derived from the fact of his occupying it and from the original act of bringing it under cultivation.[4] Inheritance resulted in the subdivision of the land and as such half of the peasantry owned less than half a hectare of land per family in 1974. The peasant paid land taxes, tributes and rendered service obligations to the state. Due to these types of exploitation the peasant lived in misery and poverty. In northern part of the country, the communal ownership was

---

3. *Ibid.*, p. 147.
4. Zoltan Gyenge, Ethiopia on the Road of Non-Capitalist Development, *op. cit.*, p. 10.

also widespread and land was owned by the community and the family members had the right to use the land. Re-allotment took place only when a family line of inheritance got broken. Every member of the family had by virtue of his descent from the founding father, a recognised and limited right to portion of his family land, but he might not usually dispose of his portion by gift or sale. Family lands, once allotted for an individual's use were usually held for life and inherited by his heirs. As rights in lands were inherited both on male and female sides, an individual had land in a number of parishes. This system had some undesirable effects like lack of continuity of farming, small size and fragmentation of holdings and frequent litigation over claims of shared land.[5]

Individual Land Tenure System

The individual land tenure systems in Ethiopia were numerous and complex and were controlled and governed by unwritten customary laws which varied from province to province. As such it would be difficult to discuss them in detail. For example in one province of Wollo, there were more than three hundred kinds of land tenures. But most of these were the same in essence and having different terms.

The most important was the Gebar system of land tenure under which a person who had acquired land by purchase, grant or inheritance, paid land tax to the government. The owner of the land was to pay land tax, asrat (tithe), education tax and wealth tax to the government. In the past the owner of the land was also subject to various dues and was required to render personal services to government officials. But these dues and services were abolished in 1944. The owner could sell, lease or mortgage his land because the right in land was inheritable. Land of which the primary interest had been vested in the church was called Semon land. The primary interest carried the right to collect and retain for church use land tax, tithe and education tax from persons settled on the land at the

5. FAO/UN, Report to the Government of Ethiopia, Land Tenure and Landlord-Tenant Relationships, Rome, 1969, p. 3.

rate laid down by law for other categories of land. In the past some members of the nobility used to transfer their gult land to the church as the tax payable to the church was simple. Some of the landowners also did the same thing in order to pay less rent but they had the right to sell or dispose of their property. But later, landowners holding Semon land paid asrat, education tax and health tax to the church. The health tax was later transferred to the government. The owner could sell their land and their rights in the land were inheritable. They did not pay land tax but had to pay for the services of the priests according to their agreement with them. But this amount was higher than the land tax.

Under the Rist-Gult land tenure system, persons having land were entitled to collect land tax from the person settled on the land at the rates fixed by law i.e., fifteen, ten and five dollars per gasha of fertile, semi-fertile and poor land respectively and at the same time to exercise judicial functions.[6] Out of the land tax collected, a uniform rate of 3.50 dollar per gasha was paid to the government for all classes of land. Rights in land were inheritable. Under Siso land tenure, out of the land possessed by the balabats (first settlers) two-thirds were taken by the government. The remaining one-third of land with the balabat was his Siso. In some cases balabat was allowed to retain one-fourth of his previous holdings. The remaining three-fourths was taken by the government. The government allowed the balabat to retain his portion (one-third or one-fourth of land) by imposing a nominal tax until it could be developed. The tax which was paid in kind varied according to the procedure of the area. The system and rates of land taxation on Siso land were the same as the Rist-Gult. The main difference between the Rist-Gult and the Siso forms of land tenure lied in their origin. While the right of Siso originated in respect of the one-third portion remaining with balabat in the manner described above, Rist-Gult rights had been granted by the emperor subsequently to members of the royal family and others as reward for meritorious services.

---

6. *Ibid.*, p. 5.

The Bill for the abolition of Rist-Gult and Siso was passed in 1964, but the proclamation was published in 1966. Under this proclamation, the land owners who held land under Rist-Gult or Siso were supposed to pay land tax as well as education and health tax on the Gebar land. On gebarless land the Gult owner was required to pay land tax at the nominal rate of fifteen, ten and five dollars for fertile, semi-fertile and poor land per gasha respectively and not at the reduced rate as 3.50 dollars per gasha for all classes of land as was the case before the proclamation was issued.

Maderia land is government land granted to government employees instead of salary for the period of office or for life. A Maderia holder had to pay asrat, health tax and education tax on his holding. He could collect his reward in kind from persons settled on his holding. This land since belonged to the government, could be transferred to other persons if the Maderia holder was convicted of a crime or transferred to another job. The holder could not dispose of his holding by gift or sale as holding was not inherited. But in 1956, the government allowed Maderia holders to get upto one gasha of Maderia land converted into Rist on payment of a registration fee. In 1959, it allowed certain types of its employees to have upto four gasha of land converted into Rist.[7]

Gult land tenure means the right to own land without paying land tax (gebar) thereon. Gult means an untaxed, but delimited area of land possessed by a person of good merit or a member of the royal family who had the right to exercise judicial power in his gult area. Land owners within his jurisdiction were liable to pay tax to and receive orders from him. Gult holders were required to give services to the government in time of war and in administering their respective zones and discharging other official duties. This Gult system of land tenure was abolished in 1966.

Lands registered as government property were called Mengist. Grants of land were made to people out of Mengist land. The remaining Mengist lands were leased to people for grazing or farming purposes. Lands taken over by the government due

---

7. *Ibid.* p. 5.

to non-payment of land tax by the landlord was called Gebratel land. This land was also leased to people by the government for farming and grazing purposes. The defaulter had the right to get back the land by paying double the amount of tax due unless it had already been given to some one else as a grant.

TABLE IV.1

Percentage Distribution of Measured Lands by System of Land Tenure in Some Provinces

Province	Gebar	Semon	Maderia	Rist Gult	Siso	Mengist	Church Gult
Shewa	53.70	13.60	3.60	5.40	9.80	12.20	0.30
Arusi	41.20	23.50	2.60	3.60	11.70	1.10	0.01
Wolega	44.33	4.9	3.69	0.92	13.70	33.13	0.07
Gemu-Gofa	16.70	5.38	—	—	0.06	77.86	0.04
Welo	81.80	13.60	—	0.10	—	—	—
Sidama	42.70	3.30	2.10	2.70	1.30	47.40	—

Source : FAO/UN, Report to the Government of Ethiopia, Land Tenure and Landlord-Tenant Relationships, Rome, 1969, p. 6.

The percentage distribution of measured land by the system of land tenure in some of the provinces in Ethiopia is shown in Table IV.1. Gebar and Mengist were the main systems of land tenure which prevailed in Ethiopia before the Revolution. In Welo and Shewa provinces, about 82 per cent and 54 per cent of the land was under Gebar. This system was also important in Wolega and Arusi provinces where the land under Gebar was between 40 and 44 per cent. On the other hand, Mengist system occupied maximum land in Gemu-Gofa and Sidama provinccs i.e. 78 per cent and 48 per cent respectively.

Landlord-Tenant Relationships

Before the Revolution, tenancy was widely practised in Ethiopia. The Land Policy Report estimated that two out of three Ethiopian farmers were tenants.[8] In some provinces like Tigre, Begemdir and Gojam, the problem of tenancy was not serious because land was owned by communities. The position of these tenants was very weak in relation to the landlords. They paid the landlord the greater part of their produce by way of rent. This amount was not fixed or controlled by any universal rental system. It varied from one-fourth to three-fourths of the produce and was determined by the customs and traditions of the region and the agreement between the owner and the tenant. In most of the cases the landlord took more than the amount fixed and on such occasions the peasants were helpless as they were not in any way organised to defend their interests. The Civil Code addressed itself to this issue by stating that the share of the produce to the landlords "may in no case exceed three-fourths".

The large landed proprietors were for the most part absentees. They resided in the capital and other urban areas and visited their estates mainly during the harvest season to collect rent and credit. In general, they entrusted their affairs to agents who exploited the tenants mercilessly. Although by law the tenant was free of all personal service obligations, tenants in many parts of the country were required to perform services such as transporting grains to the landlord or agent's home, building houses and granaries and helping in marriages and other activities. According to the available information, except Gemu-Gofa, more than 25 per cent of the owners in other provinces were absentees. In the case of Shewa province, 34.5 per cent of land owners were absentees and they owned about 44.5 per cent of the land. In Gemu-Gofa, the percentage of absentees was of course only 9.9 per cent but they owned about 41.5 per cent of the land.[9]

The percentage distribution of total area cropped by tenure in some of the province shows that more than 60 per cent area was owned holdings in Sidama and Welo. But in other pro-

8. Land Policy Project, Tenancy Reforms in Ethiopia, Addis Ababa, 1964 p. 12.
9. FAO/UN, Report to the Government of Ethiopia, Land Tenure and Landlord-Tenant Relationships, *op. cit.*, p. 19.

vince, maximum area was in rented or mixed holdings. Share cropping was the most common rented arrangement in the country. Under this system, the tenant was responsible for all the production costs while the landowner provided land. The crop was to be distributed equally. This type of rental arrangement had adverse effect on the productivity of land. The tenant would not spend on additional labour and capital unless he expected an increase in his product valued at more than double as half of the additional product would go to the landlord.[10] Moreover, the tenant had no security and permanent right over the use of land. So he had no significant "incentive to maintain or improve the fertility of the land or to make other improvements for which he will not be compensated when his tenancy is terminated."[11]

TABLE IV.2

Percentage Distribution of Total Cropped Area by Tenure in Some Provinces

Province	Owned	Rented	Mixed
Shewa	28	55	17
Arusi	38	50	12
Welega	46	45	9
Gemu-Gofa	49	47	4
Welo	61	14	25
Sidama	64	35	1

Source : FAO/UN., Report to the Government of Ethiopia, Land Tenure and Landlord-Tenant Relationships, 1969., p. 18.

Land Taxation

In the past land tax, tithe and other dues were separately paid in kind. In 1944 the land tax law consolidated all the different taxes into a fixed 'single land tax' payable in cash.

---

10. *Ibid.*, p. 20.
11. International Bank for Reconstruction and Development/IDA, The Economy of Ethiopia, Vol. III—Agriculture, Africa Development, 1967, p. 49.

But even then the land tax was not standardised and was assessed and paid in different ways. In the provinces of Tigre, Begemdir, Gojam and two sub-provinces of Welo, land tax was a fixed tribute. In these provinces land was measured and individual peasant's tax was fixed by estimating the area and fertility of his holding, or on the basis of his wealth and number of ploughs. In other provinces land was mostly measured, though not accurately, and categorised into fertile, semi-fertile and poor. The taxes were assessed on the size and fertility of the land. But the unmeasured lands in these provinces paid a consolidated tribute. The pastoral people paid tax in the form of cattle-tax and was based on the number of livestock owned. The tenants had to pay a fixed amount of his produce as tithe to the landowner who was supposed to pay the land tax in aggregate. This system was defective because the classification of land was unrealistic and unfair. A gasha of fertile land (measured) was charged at 35 dollar. In some places, the fertile land was classified as poor or semi-fertile. Further, the tax assessment did not take into account the crops grown and the number of harvests. The method of collection of tax on non-measured land was responsible for the low receipts in terms of revenue for the government. It was estimated that the taxable capacity of land could be tripled or quadrupled if a uniform cadestral system was carried out.[12]

The above analysis clearly shows that before the present military government deposed the emperor, about 80 per cent of the land was owned by four groups : the imperial family (15 per cent), the church (20 per cent), aristocracy (20 per cent) and the government (25 per cent). The church, the government and the emperor paid no taxes, while that for large landowners was nominal.[13] More than half of the Ethiopia's farmers were tenants who on the basis of either custom or law were required to pay tithe to their landowners of upto three-fourths of their total produce. The tenants were also supposed to

12. Ministry of Planning and Development (IEG), Agriculture, Wildlife, Forestry and Fishing, Documentation No. Ag—G—16, p. 26.
13. Alan C.G. Best and Harm J. de Blij, African Survey, John Willy and Sons, London, 1977, p. 490.

provide labour and other services to the landowners. Further, there existed a large number of land tenures and land taxes.

Policy for Land Reforms

The Government of Ethiopia had long been concerned with the problems of land tenure system. Even prior to the Italian invasion some important land reform laws were enacted concerning land measurement, abolition of all personal service obligations and establishing a single tax. For example, Land Measurement Law was passed in 1930 to introduce a uniform land measurement and a rational taxation system throughout the country. The law also clarified that if the land was found in excess but such land was fully utilised, the holder could continue to own it provided the holder had paid the required amount of tax. The purpose of this law was to extract more revenue and at the same time to introduce redistribution of land. The unutilised land was divided into four quarters and allotted a quarter each to the former owner, the tenants, the balabat (chief) and landless persons.

The second important law was passed to abolish all the personal service obligations. No tenant was required to perform any sort of service for the landlord. It was intended to give the cultivators more time to spend in his own work so as to provide the population with more food.[14] In 1935, the third law was enacted which established a "single land tax" to replace the petty taxes and service obligations levied on gebar (free peasant) for gasha per year. The main purpose of this law was to modernise the tax system by substituting payment in kind with cash.[15]

But there was a strong opposition from the upper class people who were to implement these laws. They regarded these laws alien to their traditions. The land measurement programme was implemented only in those regions which were inhabited by dissident groups. The upper class people also opposed the law which abolished personal services. But after some time

14. Balamberas Mahteme Sellasie, Ethiopian Observer, No. 9, 1969, p. 296.
15. *Ibid.*, p. 296.

the majority of the tenants became free from all these services. Due to tough resistance from the landlord, the single law was implemented after the Italian occupations. The first step which the emperor took after the occupation was to distribute the land to those who had helped him during the war period. All patriots were given the right to government land. Permanent ownership to holdings was granted to those who performed military services to him. From this discussion one could conclude that before the invasion, the objective of land laws was to collect sufficient revenue and after the invasion, the laws were enacted just to reward services rendered by others. It was never though at that time to pass laws for the distribution of land for the welfare of the peasants.

It was only in 1961 when the government appointed a Land Reform Committee to solve the land problems in a systematic way. The emperor made a statement about the objectives of land reform in Ethiopia. "The fundamental obstacle to the realisation of Ethiopia's agricultural potential has been, simply stated, lack of security in thc land. The fruits of the farmer's labour must be enjoyed by him whose toil has produced the crop. The essence of land reform is, while fully respecting the principle of private ownership, that landless people must have opportunity to possess their own land, that the position of the tenant farmers must be improved and that the system of taxation applying to land holdings must be the same for all."[16]

The Second Plan also fully recognised the importance of land reform for the development of agriculture. It stated: "The execution of Land Reform with the final aim of realising the proclaimed principle that 'the fruits of the farmer's labour must be enjoyed by him' should be considered as a composite part of the measures for the development of peasant agriculture. The execution of the first phase of land reform during the Second Plan is, in fact, a sine qua non for the successful application of other measures."[17]

16. Quoted from International Bank for Reconstruction and Development/IDA, The Economy of Epthiopia, Vol. II-Agriculture, *op. cit.*, p. 50.
17. Imperial Ethiopian Government, Second Five Year Development Plan, *op. cit.*, p. 119.

The Land Reform Committee which was appointed by the government to study the land tenure systems prevailing in the country and prepare a programme of land reform, divided the various aspects of land reform into short and long-term programmes.[18] The short-term programmes were : to introduce tenancy legislation to fix maximum rents and give the tenants security of tenure ; to introduce land tenure legislation and to abolish antiquated forms of tenure such as Rist-Gult and also Siso-Gult ; to introduce land tax reform including income tax on agricultural rents ; to introduce land survey and land registration reform. The long-term programmes were ; to introduce ceilings for individual land holdings and to decide the expropriation and redistribution of surplus land and compensating landlords for them ; to decide on the aid to be given to new peasant owners in the fields of agricultural credit, extension services, community development and marketing ; to study the possibility of introducing minimum holdings; to study the problems of converting communal to individual tenure ; and to study the problems of nomadic lands.

All the short-term items of the programme were included in the Second Plan which states, "in the course of the next five years, we must concentrate on the following aspects of land reforms, (a) improvement of landlord-tenant relationships, (b) progressive taxation, (c) abolition of out-dated holdings, and (d) cadastral surveys and land registration.[19] On the occasion of the release of the Second Plan, the emperor "ordered that a bill be prepared and prescribed to the Parliament for the abolition of Rist-Gult and Siso-Gult of tenure and for the improvement of landlord-tenant relationships."[20] For the successful implementation of land reform, the Second Plan proposed the establishment of government agency to prepare necessary legislation on landlord-tenant relationships, land taxation, tenure reforms and cadastral surveys. Initially, the

18. FAO/UN, Report to the Government of Ethiopia, Land Tenure and Landlord-Tenant Relationships, *op. cit.*, p. 7.
19. Imperial Ethiopian Government, Second Five Year Development Plan, *op. cit.*, p. 327.
20. Quoted from FAO/UN, Report to the Government of Ethiopia, Land Tenure and Landlord-Tenant Relationships, *op. cit.*, p. 8.

Land Reform and Development Authority was established in 1965. But later, the government had demonstrated its desire to carry out land reforms more sincerely and consequently a Ministry of Land Reform and Administration with all powers was established in 1966. This was an important landmark in the history of land reform in Ethiopia.

The improvement of landlord-tenant relationships generally includes rent control, security of tenancy and legalising leasing arrangements. "Existing trends indicate that the rent and sharing arrangements do not encourage improvements of farming methods, nor investment, and this has resulted in the preservation of a very low productivity of those farms. It is, therefore, necessary to undertake pilot studies and carefully analyse different regions on the basis of which rent should be determined and made legally binding on both landlords and tenants."[21] Inspite of the fact that two main outstanding measures, i.e., a written agreement between landlords and tenants and a fixed rate of rent payment, were neither expected to involve high cost nor liable to face strong landlord resistance, the policy measures remained largely unfilled. Most of the studies were completed on the common pattern of rental practices during the Second Plan period, but the recommendations of these studies were never implemented.

Cadastral survey was considered to be a basic condition for carrying out the overall policy of land reform. "The taxable capacity of land even under the present system of land taxation could be tripled or quadrupled if a uniform cadastral survey is carried out. The survey is planned as a long term undertaking and is intended that complete information about any piece of land should be clearly known and officially registered. The survey will also increase the availability of potential and assessible lands."[22] Although the importance of cadastral survey was realised, this could not be done during the Second Plan period. The survey was of course a long term project and cost, delays and other difficulties were bound to come, but a

21. Imperial Ethiopian Government, Second Five Year Development Plan, *op. cit.*, p. 328.

22. *Ibid.*, p. 329.

start could have been made. Without this survey, the implementation of other measures for agrarian reform would be greatly handicapped.

One of the policy measures which was considered important as a sound approach to land distribution was progressive taxation. Despite the intended introduction of this measure, the collection of land revenue was not very systematic. However, the Second Plan stated that the "analysis and setting up of the progressive model of the new tax has to be based upon careful studies of the existing taxes and their implications."[23] But due to insufficient studies carried out on existing taxes and size of holdings by different owners, the intended progressive taxation had not been given due consideration. Moreover, the late establishment of the Ministry of Land Reform and Administration was one of the main causes for the delay of the whole programme. This not only blocked the programme of land distribution, but also the possible increase of revenue to the government.[24]

As regards the abolition of out-of-date holdings, the Plan had given much serious consideration to this problem and its effects. "Owing to the large variety of existing out-of-date ownership systems the methods and procedures for abolishing them should be adjusted to the local circumstances."[25] As has already been mentioned, insufficient studies on local circumstances remained main drawback. Some studies on the multiplicity of land tenure systems had been completed, but a great portion of work still remained to be done during the Second Plan period.

The only land reform legislation passed during the Second Plan period related to the abolition of Rist-Gult and Siso-Gult by Proclamation of March 7, 1966. Under this Proclamation land owners (Gebar) who held land within Rist-Gult and Siso-Gult were supposed to pay land tax as well as education tax and health tax directly to the government treasury. The

23. *Ibid.*, p. 328.
24. Ministry of Planning and Development, Agriculture, Wildlife, Forestry, Fishing, *op. cit.*, p. 33.
25. Imperial Ethiopian Government, Second Five Year Development Plan, *op. cit.*, p. 328.

nominal rates were fifteen, ten and five dollars per fertile, semi-fertile and poor gasha of land respectively for all classes of land. The government of course submitted a Land Reform Bill in 1965 to a landlord-dominated Parliament. The bill proposed the introduction of a uniform rental system and the regulation and improvement of landlord-tenant relations. To become a law, a bill had to be approved by the two Houses, the Senate and the Chamber of Deputies. Because of vested interests the reaction of the parliamentarians to the bill had been negative.

So the land reform programmes during the Second Plan could not be implemented due to two important reasons. Firstly, the necessary studies on the existing pattern of land tenure systems could not be completed for all the provinces and consequently a uniform pattern of tenure could not be evolved. Secondly, the late establishment of the Ministry of Land Reform and Administration caused a considerable delay in the implementation of the programmes. But during the Third Plan period, the Ministry of Land Reform and Administration was assigned some special tasks.[26] "Because of the absence of cadastral maps and registration of rights in land, ownership and tenancy rights are not clear, and boundaries are not clearly demarcated. As a result, land titles are not secure. Litigation is a common problem. The Ministry of Land Reform and Administration will therefore submit to parliament legislation calling for cadastral survey, title registration and adjudication procedure. In the TFYP upto 6 million hectares of the provinces of Shewa and Arusi will be covered.

"Another urgent land reform measure to be implemented in the TFYP is the enactment and execution of legislation concerning landlord-tenant relationships, with the main aim of protecting tenants against arbitrary eviction and replacing the sharing cropping system of paying rent with a fixed rent system, but it should also provide for written leases, compensation for improvements made by the tenants, a right of pre-emption of the tenant by the landlord, and other measures designed to increase the security of the tenant. The Ministry of

26. Imperial Ethiopian Government, Third Five Year Development Plan, 1968-73, *op. cit.*, pp. 196-197.

Land Reform and Administration will shortly submit to parliament legislation to provide for the regulation of agricultural tenancy relationships.

"One of the major problems of structure that can be dealt with successfully in the TFYP period is the existence of excessively large unutilised holdings. Many of the potentially productive lands in the Empire are now held by a few persons. A progressive land tax would force these persons to put the land into production i.e., act as a negative incentive in order to avoid the burden of the tax, or the loss of control of the land through being compelled to lease or sell it to other persons who are willing to cultivate it. The Ministry of Land Reform and Administration will submit legislation to provide for a tax on unutilised land."

Over and above these tasks, the Ministry of Land Reform and Administration was also supposed to undertake studies, to device land classification criteria and formulate a tenure pattern which would help the farmers to make transactions in land. The Ministry would take all steps necessary to grant ownership rights to tenants of government land, particularly where tenants were otherwise landless.

The Department of Land Tenure of the Ministry took up that responsibility and in its Annual Report of 1968-69, the Department stated that, "the work planned for the year 1969-70 consists of completion of general land-tenure surveys and completion of the various provincial reports in a meaningful way, starting of a detailed survey of land lord-tenant relationship in different parts of the country, preparation and launching of studies on communal areas, analysis of all the available information on size-distribution of holdings and execution of more case studies of land disputes from those provinces which had not yet been covered."[27] On the basis of studies conducted, the Ministry of Land Reform and Administration prepared three important bills and submitted to the government. They were as follows :[28]

27. Department of Land Tenure, Ministry of Land Reform and Administration, Annual Report, 1968-69, Addis Ababa, p. 17.
28. The Technical Agency, IEG, An Agricultural Credit Program for Ethiopia, Stanford Research Institute, California Report No. 9, 1969, pp. 35-37.

1. *Regulation of Agricultural Tenancy Relationships*—The proposed legislation's objectives were to raise the level of agricultural development, to increase the security of the tenant on the land he was farming and to encourage investment in agriculture.
2. *Registration of Immoveable Property*—The purpose of this Proclamation was to establish a procedure for recording land ownership by a cadastral survey and, in those areas where there were disputed claims of land ownership, to settle the question through an adjudication officer or committee. It was expected that through the survey, land might be discovered without ownership claims, in which case it would become government land ultimately available for development purposes.
3. *Tax on Unutilised Land*—The purpose of this legislation was to require large landowners (20 hectares or more) to develop their unutilised agricultural holdings within a required period or be subject to a special tax over a five year period. The tax would be sufficiently large so that, in most cases, it would require either compliance with the law, or transfer of land to persons not under the owner's control, including government. The proposed legislation was also to be applied to land owners whose combined holdings throughout the empire were in excess of a designated number of hectaes.

The proposed legislation for the Third Plan period was very comprehensive in dealing with the different aspects of the tenancy. Over and above the gradual abolition of share cropping, introduction of controlled and fixed rent was to give security of tenure to tenants, to provide compensation to tenants for improvements made on the holdings in case of termination of lease, and to encourage written agreements. It was also thought at that time that these proclamations would be an important step towards advancing the welfare of the Ethiopian rural population.

But such measures were considered revolutionary by the upper class people. The success for them was questionable.

The authorities who were to execute these laws were the very people the laws were to affect. The spirit of the laws came into conflict with their personal, family and class interests. The implementation of these laws would mean the reduction of their income and limitation of power. The elite considered these laws alien to the Ethiopian tradition and they frustrated the implementation of those laws.

The Ministry of Land Reform had also put forward a draft proclamation for agricultural tenancy for the approval of the New Parliament in 1973 as it had done every year since 1968. The proclamation allowed for leases to be written or oral, though it specified that they must be written at the request of either party or at the demand of the Ministry of Land Reform. The rents to be paid either in cash or kind were to be limited to one-third of the yield in the case of share cropping. If the landlord contributed seeds, oxen and other inputs, he might obtain more provided that the total does not exceed one half of the production. The Ministry of Land Reform had the right to reduce rent if and when it might be necessary.

The most important aspect of the proclamation was that tenancies were to be terminated at the agreement of both parties or by the initiative of either. But in the case of landlord, he was supposed to prove default for taking any such decision. The initiation of mechanisation of farms was also considered as an important cause for the termination. But the tenant had to be given 2 years notice and be paid compensation for not less than two years' rent. The compensation was required for improvement whatever the reason for termination of tenancy might be.

Regarding the implementation of such measures, it was decided to appoint tenancy committee at the district level to hear appeals from the tenancy officers' decisions and arbitrate on disputes. These committees were to have equal number of tenants and landowners and be under the chairmanship of an owner-farmer. But the majority of members of Parliament regarded it too radical and it was not acceptable to them. In the years since the draft proclamation was first put forward, the landlords had ample time to prepare for it evicting any tenant that they wanted out. There is considerable evidence to suggest that during 1970-74, evictions for this reason had been steadily increasing.

Hardships caused by the tenurial system were further aggravated by the severe drought of 1969-73. The authorities failed miserably to resolve the agrarian crisis. That was why, the demand for a radical land reform for the transfer of land to those who till it became basic national slogan in the course of the revolutionary events that started in early 1974.

CHAPTER FIVE

Agricultural Credit

THE IMPORTANCE of agricultural credit, which is of an institutional character, is being increasingly appreciated in all the countries. In the economically more developed countries where agriculture is becoming more capital-intensive there is a parallel increase in the need for credit to finance more improved techniques.[1] In less developed countries the need for credit may be even more urgent. Farmers are almost without liquid resources and credit on reasonable terms is required to finance the preparation for the next crop, to make a modest start in adopting methods of farming, and often to enable them to avoid selling crops at the very low prices ruling immediately after the harvest. Further, with the implementation of rural reform programmes, the increased marketing of crops by peasants and the use of commercial fertilisers and other farm inputs, the need for agricultural credit has increased rapidly.[2]

1. FAO, The State of Food and Agriculture, Rome, 1961, p. 59.
2. United Nations, World Economic Survey, Parts I and II, New York, 1967, p.65.

The importance of agricultural credit for development purposes was recognised during the pre-revolutionary period. "New and rich agricultural lands in Ethiopia's lowlands offer opportunities for development through the resettlement of subsistence farmers living in densely populated highland areas... The depleted agricultural areas in the highlands may be restored to yield higher rates of return than at present through the introduction of more modern farming methods, including the proper use of fertilisers. Private and government funds... are needed to supplement the programme."[3]

While there was a large demand for agricultural loans in the Ethiopian rural economy, it may be said that for all practical purposes agricultural loans were not available in sufficient number as the major credit institutions did not reach the level of small farmers. The loan requirements by the rural population for making improvements in agriculture had been analysed by the Stanford Research Institute which divided the agricultural population into four general classes.[4] Class I consisted of the subsistence farmers working on small areas generally less than two hectares with about 87 per cent of the agricultural population Class II farmers were those who were in a position to cultivate two hectares of land that could be developed into economically self-supporting units but who might also be subsistence farmers. Class III were in a position to cultivate larger areas of land. They had a higher degree of education and desire for credit to make improvements in agriculture. Most of the farmers of class IV might be absentee farm holders. They were educated and owned large tracts of land suitable for plantation development projects. Their financial requirements were very large and could be met only by big institutions.

The majority of the population was in the first two categories. "It represents a sizeable portion of the major purchasing power. If efforts are made to improve the standard of living of these two classes, it would aid the development of other branches of the economy." Further, the vast majority of the rural

3. The Technical Agency, IEG, An Agricultural Credit Program for Ethiopia, *op. cit.*, p. 5.

4. *Ibid.*, p. 7.

population did not utilise the loans to increase productivity and family income, but rather to sustain the family or meet emergency conditions. As a result, agricultural credit had not usually led to a cumulative increase in productive power. The credit had remained static. After paying interest and repaying debt, the assets of the farmer, and his capacity to produce and maintain levels of consumption for himself and his family remained unchanged.[5] This normally implied that credit cycle had to be repeated to maintain his position. So the requirement to promote growth was not only to increase the amount available but also to integrate credit with assistance in applying new techniques and promoting better farm management.[6]

Sources of Credit

There were three important agencies which supplied loans to the agricultural population. These were private money lenders, banks and co-operatives. Initially, the private money lenders might have not been real money lenders, but lenders in terms of seeds, cereals and so on. Later, they supplied commodities with money. The main characteristic of those money lenders was that they did not act on a collective basis. The only collective money lenders in Ethiopia were the churches. They had a large area of land and possessed surplus products which they could lend to farmers around their holdings for the purchase of seeds and meeting their family expenses. The remaining groups of money lenders were individuals and the most important of which were the village shopkeepers who provided loans for day to day purchases at a high rate of interest, the landlords who after borrowing from the government or other agencies used to lend to tenants against the security of crops, and the private money lenders who had income in excess of their needs.

But the money lenders were not professional in the sense that they did not wholly depend on the income from interest.

5. FAO, Agricultural Credit in Economically Underdeveloped Countries, 1959, Quoted by Agency for International Development Research Review, Vol. I, No.I, 1962, p. 41.

6. *Ibid.* p. 41.

They were holding a special position in the rural economy because other agencies had offered little competition to these private money lenders due to three important reasons. Firstly, other agencies were wholesale dealers in large loans, whereas the rural population required retail credit. The private money lenders could very easily fulfil the demands and they could lend as small amount as was required by the small farmers. Secondly, the procedure followed by the money lenders at the time of lending was very simple. They did not conduct investigation as was generally done by other agencies. Since a money lender was a resident of that village, he knew all the farmers and the prevailing conditions in the village. Thirdly. as the other agencies were interested in providing medium and long term loans, the village money lenders made available loans for short period to the farmers. Consequently, the supply of loans by the money lenders was the main source for the farmers. There is no extensive research done so far and it is difficult to state with certainty the role played by these money lenders in the economic life of the rural population.

The private money lenders charged very high rate of interest. In a survey conducted by a student of the University, it was observed that about 19 per cent of the people paid more than 100 per cent interest and the remaining paid between 24 and 100 per cent.[7] The position was more serious when the money lender was also the landlord of the tenant. In that case the money lender had a double hold over the tenant. The tenant had to pay both the rent and the interest and his share in the harvested crop remained very little after meeting all the expenses.[8] This is the reason that one of the United Nations experts stated :

> "Shortage of credit may be both cause and effect of poverty. It may be a cause in that lack of ready money in the hands of farmers and merchants prevents investment in the farm and market. But it may also be the effect or symptom of poverty, in that high interest rates and high burden of debt may reflect a chronic

---

7. *Ibid.*, p. 28.
8. *Ibid.*, p. 23.

insufficiency of farmers income and permanent tendency for consumption to outrun production."[9]

2. Institutional Credit—Development Bank of Ethiopia

Agricultural financing on an institutional basis is reported to have been started in Ethiopia with the establishment of the Agricultural Bank of Ethiopia in 1945. The main activity of this Bank was to grant small agricultural loans upto Eth. $ 2,000 to farmers who had lost their cattle during the Italian occupation.[10] In 1949 the name was changed to the Agricultural and Commercial Bank of Ethiopia. During the history of the Agricultural Bank (1945-49) and its successor Agricultural and Commercial Bank (1949-51) about 740 small loans were made worth Eth. $ 781,150 and were restricted to the province of Shoa.[11] The repayment of these loans stretched over a period of three years. The interest rates charged by the Bank were 3½ per cent to 5 per cent depending upon the size of the loans. Loans were granted with proper investigation so that farmers did not divert them to non-agricultural purposes.

The Development Bank of Ethiopia was established in 1951 absorbing the former Agricultural and Commercial Bank of Ethiopia to assist in the development of industrial and agricultural production and to foster the invest ment of private capital for productive purposes. The proclamation that established the Development Bank of Ethiopia provided that the Bank "can make, guarantee or participate in loans to any industrial, agricultural or business enterprise on such terms and conditions as it shall deem appropriate." As a rule the Bank was to finance upto 73 per cent of the total cost of agricultural projects subject to adequate security. The loan term was from one to five years. But loans for longer and shorter period were also granted.

9. United Nations, Land Reform, Department of Economic Affair, New York, 1951, p. 37.
10. H.P. Huffnegal, Agriculture in Ethiopia, *op. cit.*, p. 460.
11. Statistics and Research Department, DBE, Development Bank of Ethiopia's Twelve Years Programme, Ethiopia—Trade and Economic Review, 1956-66, p. 73.

Agricultural Loans

During 1951-1967 the Bank made 3,978 loans valued at Eth. $ 42,403,000. The percentage of industrial loan was 56, agricultural loans 40 per cent and the remaining 3 per cent for other loans.[12] This position was maintained in the subsequent years. Agricultural loans were of three types—general agricultural loans, coffee loans and small agricultural loans. The agricultural loans other than coffee loans were granted to big land owners for the production of cereals, oilseeds and fruits. The reason for making advances to big land owners was that there were good potential possibilities for the development of big agricultural estates. Inspite of the fact that the effective demand for these loans was limited due to lack of good management and land tenure problems, the agricultural loans were increasing steadily.[13] The Bank for the first time financed major agricultural projects in 1957. The Matahara Plantation Estate received a loan of Eth. $ 65,000 for the purchase of machinery. Another Eth. $ 300,000 was granted for the same project in 1958. The Bank had also played an active role in the development and extension of the first large agricultural enterprise in the Awash Valley where rice among other crops had been cultivated for the first time in Ethiopia. Another factor contributing to the growth of agricultural loans was the purchase of tractors and other mechanical equipment. There was some fall in the number of loans in 1961 which was due to the scrutiny of applications but from 1962 onwards these loans were again growing steadily. Out of the total loans the share of agricultural loans (general) sanctioned from 1951 to 1968 was 50 per cent.[14]

Coffee Loans

The Bank had really played an important role for the development of the coffee industry. In 1953 the Bank decided

---

12. Technical Agency, IEG, An Agricultural Credit Program for Ethiopia, *op. cit.*, p. 185.
13. Statistics and Research Department, Development Bank of Ethiopia's Twelve Years Programme, *op. cit.*, p. 75.
14. Technical Agency, IEG, An Agricultural Credit Program for Ethiopia, *op. cit.*, p. 185.

to co-operate very closely with the government in carrying out its long term development programme of bringing forest area in the south and west of Ethiopia under coffee cultivation.[15] So the first branch of the Bank was established in Jimma in that year. The Bank granted the so-called "Standard Coffee Development Loan" of Eth.$ 5,000 per person for the forest or other suitable land for coffee. In 1956 a second branch was opened in Gore. The coffee growers also helped the Bank by undertaking joint responsibility for the repayment of loans granted to members of that group. A third branch was established in Yiraglem for providing loan facilities in some of the area of Sidamo province. Later on, other branches were established in Asmara, Bale, Dese, Gojam and Nazeret.

The credit assistance started in 1953 continued to expand partly due to the high prices of coffee in the world market. In order to provide some technical assistance to some farmers, the Bank engaged the services of foreign coffee expert. The free technical assistance/guidance for coffee producers was also provided in some of the coffee growing areas. In addition, the Bank purchased and installed a coffee washing plant at Suntu in 1955. Although this plant did not prove economically successful, it had influenced the coffee growers to process their coffee in the most modern manner thus contributing to the improvement of coffee quality.[16] During the period of 1951-59 the Bank advanced coffee loans to the extent of about 19 per cent of the total loans.

In 1960 the amount of coffee loans decreased very substantially. This was partly due to the fall of coffee prices in the world market. With the result the clients were reluctant to meet the commitments. Moreover, the loans were not granted for sufficient period to get the fruits of the investment by the farmers.[17] The loans were granted for about four years. But actually coffee trees start giving fruits only after six or seven years. On the other hand, the Bank stopped advancing loans for bringing new areas under cultivation. The loans were granted

---

15. Statistics and Research Department, DBE, Development Bank of Ethiopia's Twelve Years Programme, *op. cit.*, p. 76.
16. *Ibid.*, p. 77.
17. *Ibid.*, p. 78.

only for improvement of existing farms and plantations. But even then the percentage share of coffee loans to the total loans was 23 in 1970.

Small Agricultural Loans

Although the Bank had helped big land owners and plantations, its policy for advancing loans to small agricultural farms was not encouraging. This was one of the important reasons why small farmers had to depend on private village money lenders even though they charged high interest. During 1951-60 period, the Bank made only 1,285 small agricultural loans valued at Eth. $ 1,475,000 or on an average of Eth. $ 1,150 per loan. These loans ranged from Eth. $ 500 to Eth. $ 1,500. From Table V.1 it would be clear that the maximum number of loans were granted in 1958. From that year onwards the number of loans and amount advanced had started decreasing and in 1961 the policy of making small loans was basically discontinued.[18]

TABLE V.1

Number and Amount of Small Agricultural Loans

Year	Number	Total Amount (Eth.$ 000)
1951	88	83
1952	69	86
1953	106	101
1954	108	130
1955	116	136
1956	142	168
1957	134	178
1958	312	334
1959	156	190
1960	54	69
	Total 1,285	1,475

Source : The Techincal Agency, IEG, An Agricultural Credit Program for Ethiopia, SRI, California, 1969, p.182.

18. The Technical Agency, An Agricultural Credit Program for Ethiopia, *op. cit.*, p. 183.

There had never been a public announcement that the small loans would not be granted. The small agricultural loans that were advanced were no longer shown in a special category. In 1964 the Bank issued a regulation that the minimum loans would be of Eth.$ 10,000. This policy raised so much criticism that the minimum limit was abolished. It was said that the discontinuation of these small agricultural loans was mainly due to the problem arising through the collection of principal and interest. Secondly, even the rate of interest which was 7 per cent charged from the small agriculturists was too high for the borrowers to bear, but it was uneconomical for the Bank since a rather heavy establishment had to be maintained. Thirdly, there was no system of supervised agricultural credit and many of the loans were reported to have been used for non-agricultural and non-productive purposes. As a result, at the end of 1970 about 224 small loans were outstanding amounting to Eth.$ 200,000 or an average of Eth.$ 890. Many of these loans had been reduced to very small amounts, even under Eth. $ 100 and in all probability would have been written off by the Bank.[19]

The past history of the programme of the Bank in respect of small agricultural loans and the reasons for the curtailment of loans were of such a nature that the unsuccessful results did not stand as barrier to the advancement of these loans. "Small farmers should not be kept without help if the proper use of money and its contribution to an economic development even on a modest scale, can be secured."[20] With this idea in mind, the Bank planned to help the small agriculturists through the co-operatives. It granted loans to these societies at a reasonable rate of interest. But the success of these co-operatives was not satisfactory due to lack of funds. For example, the Bank made only 5 loans to these societies in 1967. The Bank also made an agreement with the Addis Ababa Bank to provide loans to these co-operatives. But the Addis Ababa Bank was authorised to charge 9 per cent rate

19. *Ibid.*, p. 183.
20, Development Bank of Ethiopia, Annual Report for 1961, Addis Ababa, p. 2.

of interest plus 1.5 per cent for guarantee. If the commission of co-operatives for handling charges is also added, the interest paid by the members of the co-operatives would have ranged from 12.5 to 13 per cent which the small farmers could not afford.[21] So the Bank was not able to help the small farmers who were the major part of the rural population.

Institutional Credit—Other Banks

The Commerical Bank of Ethiopia started its operations on January 1, 1964 to carry out all types of banking business and operation and also make short-term and medium-term loans. The CBE's lending programme was mainly to advance loans to industry and commerce. There was no fixed limit for minimum loans. A loan might be as low as Eth $ 300. But generally loans ranged from Eth $ 5,000 to Eth $ 10,000.

TABLE V.2

Percentage Distribution of Loan by Industry

Industry	1966	1967	1968
Industrial	20	30	23
Agricultural	5	5	9
Exports	19	16	23
Imports	18	18	17
Building & Constuc.	13	10	7
Local Trade	24	23	20
Miscellaneous	1	—	1

Source: Commerical Bank of Ethiopia, Annual Reports 1967 and 1968, Addis Ababa.

Out of the total loans advanced by the Commerical Bank only 9 per cent were granted to agriculture in 1968 though its percentage was only 5 in 1966. About 88 per cent of the total agricultural loans were for cotton, sugar and fruit plantations and only 12 per cent for other commodities. The small

21. The Technical Agency, IEG, An Agricultural Credit Program for Ethiopia, *op. cit.*, p. 188.

agriculturists were given little attention by this Bank. In 1967 the Ethiopian Farm Development Share Company was established to engage directly in farming and extend credit to farmers who were land owners. The credit was in the form of services rendered such as land clearing or in cash for the purchase of seeds, insecticides and fertilisers. This also helped the big landowners and not the small farmers.

In order to purchase and finance the sale of specific agricultural products such as wheat, oilseeds and pulses, the Cores Company was established by the Development Bank of Ethiopia in 1955. The primary function of this company was to extend credit to flour and oil mills and to exporters.[22] The credit advances ranged from 80 to 85 per cent of the market value of the commodity purchased by the Company for the mills. The interest rate charged for these advances was 10 per cent. The security for the credit consisted of the farm product stored in the guarded warehouses rented by the Company. But the Company too was limited in its activities due to its inability to receive adequate funds from the Bank and the higher rate of interest charged from the Company. Further, the Company was unable to make small credit to farmers or to provide them on credit basis with improved seeds, insecticides or fertilisers repayable in cash or kind.[23]

Instituti nal Credit—Indirect Agricultural Credit

There were also a number of sources of indirect agricultural credit in Ethiopia such as issue of seeds, fertilisers and in some cases free ploughing.[24] The Tandaho Plantation and Elberet Plantation were engaged in such activities. The Tandaho Plantation provided cotton seeds and cash for the payment of labour. When the cotton was purchased from the farmers, loans were collected. Similarly, Elberet Estate was extending credit with interest. At the time of delivery of crops to the

22. H.P. Huffnegal, Agriculture in Ethiopia, *op. cit.*, p. 463.
23. The Technical Agency, IEG, An Agricultural Credit Program for Ethiopia, *op. cit.*, p. 219.
24. *Ibid.*, p. 55.

Elberet Estate, 30 per cent of the value of crop was deducted as partial repayment of loans until the loan was fully paid.

Other important sources of indirect agricultural credit were the Ethiopian Grain Corporation, the Ethiopian Tobacco Monopoly and the Ethiopian Spice Extraction Share Company. The Ethiopian Grain Corporation was established to expand agricultural production through market stabilisation and improve the quality of the products. The Corporation also advanced loans for short period to farmers. In 1967, the Corporation successfully distributed and collected from 14 farmers of Eth. $ 40,000 worth of seeds. In 1968, the Corporation distributed seeds worth Eth. $ 100,000. The management was in favour of providing loans to small farmers on a small scale but due to some administrative difficulties, the Corporation only advanced loans to those who owned more than 12 hectares of land and a tractor.

The Ethiopian Tobacco Monopoly was set up to establish experimental stations to develop improved seeds and buy tobacco from the farmers at a fixed rate set one month before the beginning of the crop season. Every year about 1500 of the best tobacco farmers were provided with free agricultural tools. About 40,000 farmers were involved with farms averaging less than one hectare. The purpose of the Ethiopian Spice Extraction Share Company was to promote the raising of chilli peppers and processing and export. Seeds were provided free and an amount of Eth. $ 200 was also advanced per hectare free of interest. The funds advanced were deducted from the purchase of peppers from the farmers.[25] But all these types of loans formed a very negligible part of the total loans advanced by these Companies.

3. Agricultural Co-operatives

In a developing society, a co-operative form of organisation can play a very important role in increasing production and offering all people an equal opportunity to contribute to the economic and social progress of the country.[26] This has been

25. *Ibid.*, p. 57.
26. Imperial Ethiopian Government, Second Five Year Development Plan, 1963-67, *op. cit.*, p. 153.

proved by the experiences of many developing countries. In Ethiopia, co-operation as a way of life has been and continues to be a tradition in finding the solution in socio-economic problems of the people.[27] The programme for co-operative development was formulated and had been included in the Second Five Year Development Plan (1963-67). "Having in view the importance of agriculture in the development of the national economy as a whole, during the forthcoming five-year period special attention will be devoted to the enlarging and promotion of farmers' co-operatives."[28]

One of the important tasks of farmers' co-operatives was to provide credit to farmers. "Much large financial means than the farmers have at their disposal are necessary in order to introduce the use of new implements and tools, to utilise larger machines, to spread the use of better seeds, fertilizers and chemicals for plant protection, to use better breeders, establish orchards and vineyards, etc. It will not be possible to accomplish successfully such a vast programme without the adequate organisation of agricultural credit on a co-operative basis. The co-operative is able to obtain larger credits from the bank and thus reduce costs and it can distribute credit among its members most justly."[29]

The concrete targets for these co-operatives included therein were :—

(a) establishment of 20 co-operatives at the rate of 4 each year ;

(b) Training of personnel for employment in the co-operative department ;

(c) provision of comprehensive co-operative legislation ; and

(d) provision of finance for the co-operatives.

---

27. Ministry of National Community Development, IEG, Labour, Co-operatives, Social Welfare and Community Development in Ethiopia —A Brief Review, 1957-1964, No. 1964, p. 23.

28. Imperial Ethiopian Government, Second Five Year Development Plan 1963-67, *op. cit.* p. 154.

29. *Ibid.*, p. 156.

During the Second Plan period, 20 co-operative societies were set up with altogether about 4,500 members and with a paid-up capital of over Eth. $ 200,000. The Co-operative Law, which was passed in 1966, laid down the broad framework of operation of the co-operatives. It cleared the way for government participation in co-operative societies and provided a special status to them by exempting them from income-tax. On the other hand, as there was no training facilities in the country, senior officials had to be trained abroad. But the education of members of co-operatives suffered for want of fund and training facilities.[30]

The greatest difficulty that the co-operative programme had faced was in respect of supply of credit to the societies. During the Second Plan period only Eth. $ 383,000 was made available to the co-operative societies by the Development Bank and the Commercial Bank. Consequently, few societies had been able to advance loans to their members in a significant manner. Secondly, the interest rate charged by the co-operative societies was also very high. It was 13 per cent to cover the 9 per cent interest on the Bank loan, 1.5 per cent handling charges, 1.5 per cent guarantee of the Bank and 1 per cent handling charges by the co-operatives.[31]

Co-operatives during Third Plan

Recognising that "co-operation seems to be the best way of mobilising the latent organisational and financial capacity of farmers...development planning has laid emphasis on increasing use of co-operation in a wider range of productive activities ; in view of the overwhelmingly agricultural character of the economy, its greatest use will be in the rural sector."[32] The Plan further stated that the multipurpose agricultural co-operatives was one of the best institutional forms for car-

30. Ministry of National Community Development and Social Affairs, Proposals for Third Five Year plan (1961-65 EC) 1968-73, Addis Ababa, 1968, p. 45.
31. The Technical Ageney, IEG, An Agricultural Credit Program for Ethiopia, *op. cit.*, p. 160.
32. Imperial Ethiopian Government, Third Five Year Development Plan, 1968-73, *op. cit.* p. 380,

rying out complex programmes necessary for the development of peasant agriculture.

In the Third Plan it was proposed to establish at least 300 new co-operatives. Out of these, about 250 would be multi-purpose agricultural co-operatives primarily covering credit supply and marketing.[33] To carry out the programme of co-operative development, it was also proposed to set up a National Co-operative Investment Fund which would be administered by a special co-operative section in the Development Bank of Ethiopia. It was supposed to dispose of about Eth. $ 26 million as investment loans for fixed assests and for working capital. It was expected that contribution of co-operative members to investment funds would reach about Eth. $ 5 million during the Third Plan.

But it was also made clear that the development of co-operatives among the rural population, especially if subsistence farmers were to be included, would be slow.[34] The proposed number of co-operative organisers would be under great pressure and responsibility. They had not only to enlist a sufficient number of educated members but also to maintain books of account, records of the meetings, develop marketing outlets and establish loan committees that would operate under sound policy. So the shortage of trained manpower was the major problem in developing co-operatives.

Inherent Factors

The various programmes that had been directed towards advancing agricultural loans had had varying degree of success. As far as large agricultural loans are concerned, the efforts made by different credit institutions could be regarded as successful. But the structure of the credit system as a whole was not adopted to the needs of small farmers. The money lender was holding a special position in rural Ethiopia and other credit institutions offered little competition. The importance of co-operatives to replace the private money lenders was recogni-

---

33. *Ibid.*, p. 381.
34. The Technical Agency, IEG, An Agricultural Credit Program for Ethiopia, *op. cit.*, p. 162.

sed in Ethiopia during the economic planning, but again these co-operatives had also to depend on the existing banking system for their credit needs.

Over and above the shortage of credit, there were a number of factors inherent in the Ethiopian economic structure which were responsible for not only retarding the expansion of agricultural credit but also the growth of agricultural sector as a whole. These factors were small and fragmented land holdings, landlord-tenant relationships and communal ownership of land.[35] More than 50 per cent of the holdings were less than one hectare. The farmers cultivating such small lands were basically subsistence farmers and the size of their holdings could not justify bank loans. They were also scattered over different areas which further increased the difficulty of obtaining credit from banking institutions.

The legal status of the tenant was not secure under the landlord-tenant relationships and he was subject to the will of the landlord and in most of the cases he was put off the land after the crop harvest. Under such land arrangements, it was difficult for the farmers to obtain loans from the banking institutions. Another important form of land ownership was communal under which land belonged to the extended family. The cultivation of land by a farmer was for a specified period. The demand for loan from such types of farmers was purely short period because he had to leave the land after a stipulated period.

The banking institutions had also to face a number of problems at the time of advancing loans to these small farmers. The most serious problem was the falsification of assets. At the time of the visit of the appraiser of the bank, some of the farmers might borrow equipment and cattle from their friends to show their assets. The appraiser was unable to value the farmers real worth in order to provide loans to them.[36]

Rural credit of course is not the single factor that will transform subsistence type of agriculture into modern one, but if it is a part of a co-ordinated programme, it can help to

35. *Ibid.*, p. 8.

36. *Ibid.*, p. 89.

achieve the desired goals at a rapid rate. The Ethiopian Government proposed to introduce land reform measures such as distribution of land, consolidation of small and fragmented holdings. These measures also required credit and marketing facilities. Ethiopia recognised the need for rural credit for agricultural development, but the base for its sound policy had not been laid before the Revolution.

CHAPTER SIX

Marketing Organisations

BESIDES the need to overcome structural obstacles and to supply farmers with capital, it is necessary to create conditions in which farmers will have substantial material benefits by adopting innovations. The existence of favourable market prospects and convenient marketing facilities for agricultural products are very essential for the non-monetised agricultural sector for its development.[1] This enables not only the efficient marketing of their products but also assures favourable prices.

In the most advanced sense of the word, marketing is a combination of activities by which agricultural produce and raw materials are made ready for consumption and reach the final consumer in a suitable form at the time and the place he wants them. It includes assembly, storing, transporting, grading and standardisation, financing, risk taking, etc. In economically less developed countries, a marketing link is required for "transition from a typically subsistence farming to a typical production for

1. Ministry of Planning and Development, Third Five Year Development Plan (Second Draft) *op . cit.*, p. 29.

sale'".[2] What is still common in these countries is the small holding, worked primarily to feed the farmer's family, but producing a small seasonal surplus which is sold at a villlage market to pay taxes and to meet other requirements. This semi-subsistence activity of course represents a step forward from pure subsistence activities. But it is never likely to raise an agricultural economy to a level where it can save and develop in a cumulative fashion. In other words, since the markets within the reach of the farmers' own means of transport can absorb only small surpluses, they have little or no incentive to expand production beyond the level already attained.

It stands to reason, therefore, that there must be an assurance that after the harvest, buyers will come and pay remunerative prices for the farmers' produce in order to transport them to other places. Under these circumstances, he has the necessary incentive to produce more. One of the main reasons for stressing the significance of marketing is its direct impact on production. Since a large section of the population must continue to depend for their living on agriculture, an FAO expert says, "Any plan of economic development that aims at diminishing the poverty of the agricultural population, reducing consumer food prices, earning more foreign exchange or eliminating economic waste has to pay special attention to the development of efficient marketing for food and agricultural products."[3]

Small Surpluses

During the pre-revolutionary period, the great majority of Ethiopian peasant farming families (except those who grew coffee) produced and even now producing agricultural products for their own consumption and only small surpluses of low and mixed quality were offered for the market.[4] This generally decreased the quality and price. This together with the high

2. Girma Zawdie, What Marketing Means to the Agriculturist, The Ethiopian Herald, Sept., 1969, p. 4.
3. Quoted from Girma Zawdie, What Marketing Means to the Agricuculturist, *op. cit.*, p. 4.
4. Ministry of Planning and Development, Third Five Year Development Plan (Second Draft) *op. cit.*, p. 29.

cost of transportation and lack of farm-to-market roads was the main cause for the complex chain of middlemen and high margins between producers' and consumers' prices. This reduced not only the purchasing power of the consumer but also the farmer's incentive to produce more.

The government had taken some measures to transform the marketing system from a "passive collector of surpluses to an active promoter and organiser of a market-oriented agriculture".[5] Before we discuss the various measures taken in this direction and their appraisal, it would be essential to refer in detail to the marketing system prevailing in Ethiopia during the pre-revolutionary period. Further, since three commodities such as grains, pulses and coffee contribute maximum to total agricultural production, the marketing system of these three commodities will be discussed here. Marketing is generally composed of all the activities which are required for transporting products from the farmers to the final consumers. The important activities are buying and selling, storing, transporting, grading and standardisation, financing, etc.

1. Buying and Selling

Many types of merchants were involved in buying and selling activities of grains and pulses in Ethiopia during the pre-revolutionary period. A merchant may be defined as "an entity which takes ownership of commodity with the intention of reselling at a price high enough to cover all the costs and provide, except for non-profit organisations, a profit."[6] As such six types of merchants could be identified who were dealing with sale and purchase of grains and pulses in Ethiopia. They were farmers' co-operatives, assemblers, wholesalers, retailers, exporters and millers and processors. In addition, there were also agents and brokers who were actively taking part in marketing in three rural, town and terminal markets.

---

5. Ministry of Planning and Development, Third Five Year Development Plan (Second Draft) *op. cit.*, p. 31.

6. The Technical Agency, IEG, Marketing of Grains and Pulses in Ethiopia, Stanford Research Institute, California, 1969, p. 51.

The co-operative societies which were small in number handled the purchase of the farmers' products and marketed them. They used to purchase the marketable surplus of their members after the harvest and sell the stocks at higher prices. The most important society was the Kala Duba Farmers' Co-operative near Gondar. The members of the society believed that their society had enabled them to realise higher prices for their products than they could get by selling to the merchants. For example, in 1967, when chickpeas were delivered to the society, the local price was Eth.$ 7 per quintal. After three months, the society was able to sell the chickpeas at Eth.$ 15 per quintal.[7]

The most important merchant was the assembler who played a dominant role in assembling a large amount of grains and pulses. His job was to procure directly from the producers or from other small assemblers and sell in bulk quantities to other merchants. They were mostly farmers who collected small quantities of grains and pulses from their neighbours for resale to merchants. There were also a large number of wholesalers involved in the collection of large supplies of grains and pulses in the expectation of a seasonal rise in prices. They had sufficient seasonal storage facilities. There was no problem for them in getting bank loans for this purpose. The largest single wholesaler was the Ethiopian Grain Corporation which was very active in the purchase and sale of grains. Retailers were also performing the activities of buying and selling though on a small scale. They used to sell small quantities of grains and pulses directly to consumers and processors. Although they had a low level of sales, the rate of turnover was relatively high. They usually held small inventories but the competition amomg them was very high.

Exporters with bank credit were involved in buying on large scale either from wholesalers or assembler-wholesalers. They used to accumulate the stock of grains and pulses and arrange to export them. Most of the exporters had direct contacts with their agents in other countries. They had storage facilities and cleaning machines. With the increase in the

7. *Ibid.*, p. 52.

consumption of wheat flour, the activity of flour mills also increased during the pre-revolutionary period. There were 11 commercial flour mills which were buying mostly from the Ethiopian Grain Corporation on credit and also from other wholesalers. Some mills were also buying directly from the farmers. Small flour mills were found throughout the country. The owner of the grain had to pay Eth.$ 1 to 1.70 per quintal depending on the quantity milled.[8]

Assemblers and wholesalers were also utilising the services of agents and brokers in making large purchases. Agents were advanced money and told the price at which they would be credited for their purchases. The actual commission paid to the agent was usually Eth.$ 0.5 per quintal with the agent paying the labourers to load the grains. Transportation charges were paid by the interesting party. The procurement through agents was cheaper because it did not involve other costs as establishment of office, etc. In contrast to agents, the assemblers and wholesalers also employed brokers. They never took possession of the commodities and their job was to bring buyers and sellers together. They had sufficient knowledge about the prices and other conditions prevailing and helped to perfect the operation of the marketing system. The brokerage fee was Eth.$ 0.10 per quintal throughout the country. But the interesting point is that these brokers used to greet the farmers on their arrival at the market and provide food and other necessary facilities to them until their goods were sold.[9]

So far as the classification of markets is concerned, there were three types of markets in rural areas.[10] One of them was the farmer's house where farmers used to buy from one another for resale. Another important place for sale and purchase was the roadside stalls. The idea was to provide a place for sale of commodities closer to the farmers. Generally, sale and purchase of commodities was conducted in the morning so as to

---

8. *Ibid.*, p. 57.
9. *Ibid.*, p. 59.
10. Based on the Survey conducted by the Stanford Research Institute for its Report on Marketing of Grains and Pulses in Ethiopia, 1969.

enable the farmers to go to other markets at a distance. But it was through local markets that most of the marketable surplus of grains entered the marketing system. These markets functioned not only as assembly points for surplus grains and pulses, but also as retail distribution outlets for locally produced and imported goods. These markets used to meet periodically with Saturday as the common day followed by Thursday and Tuesday. Retailers seldom had more than a cover space spread on the ground. Assemblers had to remove their purchases the same day. The storage structures were only suitable for short term storage. They were inadequately protected against theft, insects and dampness. The average distance people had to travel to market was between 6 and 12 kilometres.

There were also periodic as well as regular town markets in urban areas. These markets were similar in function as the rural markets and were both for consumers and producers. But they were large in size. In fact, these were assembling markets for terminal markets and for exports. The important town markets were at Dabre Zeyt, Nazret, Dabre Margos, Bahar Dar and Gondar. Agents played a dominant role in such markets. Merchants coming from surrounding areas used to sell their grains and pulses to these agents who sent these purchases to terminal markets. There were also a large number of reltailers and wholesalers who were engaged in a wide range of commodities. They had adequate storage facilities. But such markets were relatively insignificant in the overall movement of grains and pulses in the country.

Addis Ababa and Asmara were the two terminal markets in Ethiopia. The services of brokers were utilised by merchants for all major sales and purchases. In these two markets, the forces of supply and demand determined the prices of the commodities. Actually, the prices in these markets determined the prices of all other markets. They operated daily and constantly received supplies from merchants. Retailers were storing grains and pulses with the intention to sell at higher prices but the wholesalers were doing their business through brokers.

All the three types of markets lacked sanitation and hygienic conditions in the sale of foodstuffs. There was no provision for water and garbage disposal facilities. The construction of Mercato in Addis Ababa had been done but that market also lacked such facilities. The improved transportation and communication facilities were also not available in these markets.

Due to lack of statistical data, it is difficult to say with certainty about the percentage of population engaged in the marketing of grains and pulses. But it has been estimated that about 18 per cent of the population had depended upon the marketing activties. These estimates further show that grains on an average passed through 2.6 transactions between the producer and consumer and pulses about 3.2 transactions. Assemblers handled about 60 per cent of all marketable surplus of grains and wholesalers handled about 20 per cent of the total trade.[11] Further, it has been estimated that there were 12,500 to 25,000 grain merchants active in rural areas and 4.000 to 8,000 merchants in the towns. In Addis Ababa there were about 25 major grain dealers with combined storage capacity of about 100,000 m.t. But in the marketing chain, there was little functional specialisation and the same merchant might act as wholesaler, retailer, transporter, and financier depending upon the customer.[12]

2. Storage Facilities

The storage facilities on farms, particularly for grains were very poor and the investigations show that the losses might be as high as 25 per cent of the quantity stored.[13] The farmers had pits which could hold upto 100 quintals. In some cases the farmres stored grains for several years. But the stores offered very little protection from moisture, insects, rats and other rodents. For an efficient marketing system, storage facilities for the merchants are very essential. After the harvest, the farmers are in a hurry to sell their crops in order to meet their requirements while consumers demand these commodities

---

11. *Ibid.*, p. 67.
12. John Holmberg, Grain Marketing and Land Reform in Ethiopia, Research Report No. 41, Scandinavian Institute of African Studies, Uppsala, 1977, p. 9.
13. Ministry of Planning and Development, Third Five Year Development Plan (Second Draft), *op. cit.*, p. 55.

over a longer period. But due to lack of storage facilities, merchants suffered heavy losses of quantity stored.

Merchants in the urban areas had better storage facilities. They usually stored in their houses which were well protected from the weather although poorly ventilated. As mentioned earlier, the Ethiopian Grain Corporation was the largest to purchase grains from the sellers and had storage facilities. The Corporation was required to keep uniform grain prices all over the country throughout the year. This was done to some extent by transporting grains from the provinces where they were in plenty to areas where they were scarce and by buying up all available grains during the harvest season at a fixed price for selling at a uniform price during the rainy seasons.[14] The grain stores of the Corporation were situated at vantage points from where the grains could easily be transported to any marketing area. Before the Revolution, the Corporation possessed 29,000 tons of modern bulk storage capacity at 13 locations. Addis Ababa had the biggest store and was scientifically constructed. Its deep foundations of reinforced concrete and its thick steel walls were capable of isolating the inside of the store from outside temperature. As a result, grains could be kept in that store for years without losing their nutritional value.[15]

3. Financial Arrangements

The main source of merchants for financing their marketing activities was the Commercial Bank of Ethiopia. One of the main functions of the Bank was "to assist by making loans and advances available to firms, merchants, and traders to expand their business."[16] Before the Revolution, the Commerical Bank had about 40 branches in the country. A large amount was advanced specifically as merchandise loans whereby the Bank held title to the goods that were stored by the merchants although each party had a lock on the warehouse doors. But

---

14. Ministry of Information, Agriculture in Ethiopia, *op. cit.*, p. 27.
15. *Ibid.*, p. 78.
16. K.A. Cherian, Ethiopian To-day, An Upto-date Illustrated Review of Economic Development, *op. cit.* p. 61.

200 per cent security on each loan granted was required by all the banks. This had restricted the amount being advanced by the banks because the merchants did not have real property owned by them. Banks were advancing loans of about Eth. $ 1,000 to each merchant at interest varying between 7 and 9 per cent.[17]

The Development Bank of Ethiopia established its subsidiary the Ceres Company Limited for the purpose of making short-term seasonal loans and advances to merchants and millers for the purchase of cereals and pulses. This Company advanced money upto eight months and financed inventories upto 90 per cent.

Larger assemblers and wholesalers made most of their sales on credit with one month being the most usual repayment period. Some merchants used to buy from assemblers and wholesalers on cash. But the practice of small farmers was to sell to larger merchants on cash. Retailers used both cash and credit for their sales. In some cases, farmers also made forward sales of grains and pulses to merchants. They agreed to supply a given quantity of produce of a certain standard on a specific date and place at an agreed price.[18] Such forward sales were made against cash advanced by the merchants to the farmers. Forward sales by both farmers and merchants were common in export sales but less so in the local trade.

4. Transportation System

All marketed supplies must be transported and in fact "the availability and cost of transportation is one of the major determinants of the cost and efficiency with which a marketing system performs its functions. In addition, transportation largely determines the amount and location of the marketable surplus of agricultural production that can be economically integrated into the marketing system. The spread of a country's

---

17. The Technical Agency, IEG, Marketing of Grains and Pulses in Ethiopia, *op. cit.*, p. 82.
18. *Ibid..* pp. 82-83.

market economy is very closely associated with the expansion of transportation network."[19]

Ethiopia has two rail lines of primary importance—one from the port of Djibouti to Addis Ababa and the other from the port of Massawa to Agardot through Asmara. But these two rail lines transported only a small amount of grains and pulses during the pre-revolutionary period. Most of the agricultural commodities were transported by trucks. The roads at that time were not only insufficient to meet the country's requirements but were also in poor conditions. "The worse the road, the higher the cost of transport."[20] The consumers had to pay materially more than would be necessary with a reasonably improved road system. However, some efforts were made to improve the coverage and condition of roads and there had been a marked decrease in the freight tariffs. In 1953 the average highway transport tariffs from Addis Ababa to Aseb was 8.7 cents per ton-kilometre and it decreased to between 1.7 and 5.2 cents in 1966. Similarly, the decrease from Addis Ababa to Asmara was 9.3 cents to 3.3 cents during the same period.[21]

The animal and human porterage undertaken by the farmers and assemblers was "a major element in moving surplus grains and pulses to primary markets and hence into the marketing system."[22] Other small merchants also used this type of transport in moving their purchases from rural markets to larger markets. But this system of transportation was very costly. It was estimated at that time by the Highway Authority that the average cost of carrying goods by mules and camels was Eth. $ 2.40 ton-kilometer. This had largely restricted the availability of surplus grains and pulses.

---

19. *Ibid.*, p. 84.
20. Chester C. Burdick, Highways in Ethiopia, Ethiopia Observer, January, 1958, p. 403.
21. Imperial Highway Authority, Rates and Costs of Freight and Passenger Movement in Ethiopia, December, 1968.
22. The Technical Agency, IEG, Marketing of Grains and Pulses in Ethiopia, *op. cit.*, p. 89.

5. Grades and Standards

Commodity grades had not been formally established for grains and pulses. The varietal differences were the most recognised feature of intra-commodity differences.[23] In 1967, the Ethiopian Grain Corporation purchased wheat in Nazret on the basis of four grades, i.e. white, brown, white-brown and white-brown-black. These grades were based on a fair average quality grade. The main reason for the difference in quality and grades was the mixed nature of the Ethiopian seed stock. Pure seeds were not used anywhere and thus all the grains produced and sold were composed of mixed varieties. The marketing system was such that there was insufficient penalty for the inclusion of foreign matters with the commodity sold.

The metric system was the standard system of weights and measures. But there had been very little government checking of weights and scales used. The illiterate producers had to depend on the merchants to tell them the weight, even though the scales used indicated it. The merchants could very easily cheat the farmers. In one of the surveys it had been found that at Eth $ 18 per quintal, the farmer was cheated of Eth. $ 3.[24]

Marketing of Coffee

The marketing and processing of coffee was predominantly in the hands of private traders of which two buyers could be distinguished—middlemen and exporters. The middlemen were also co-operative societies and committees of merchants. With the help from the Ministries of Agriculture and National Community Development, the Gomma Coffee Farmers' Co-operative, established in Kefa in 1965, enabled its members to earn Eth. $ 3 more per 17 kg. of coffee than their non-members in 1968.[25] Further, the coffee sold by this co-operative society was of a better quality as well. The Darassa Coffee Farmers'

23. *Ibid.*, p. 111.
24. *Ibid.*, p. 116.
25. The National Coffee Board, The Story of the National Coffee Board of Ethiopia, 1957-68, 1969, p. 15.

Co-operative near Dilla in Sidama, the Berra Coffee Farmers' Co-operative in Sidama and the Jimma Coffee Farmers' Co-operative in Kefa had also shown considerable progress in the marketing of coffee.

In the marketing centres in the interior, committees composed of coffee merchants were formed. Their job was to buy coffee from the farmers and sell it to exporters. The coffee was checked by the inspectors with the standard set by the National Coffee Board. The coffee was then sent to the Coffee Inspection Centre in Addis Ababa with the inspector's signature of approval. Upon mutual agreement among coffee merchants in certain committees, offender was fined by the committee to save time going to court and thereby facilitate the coffee trade.[26] At that time there were well over 40 such committees in the coffee growing provinces of Ethiopia. In 1963, the Ethiopian Coffee Exporters' Association was formed with the object of attaining a better understanding among exporters and aimed at helping the development of the coffee trade as well as acting as liaison between exporters and the National Coffee Board.

Upgraded Quality

The main difference between the marketing of grains and pulses on the one hand and coffee on the other was that nothing much was done to improve the quality of grains and pulses. The marketing system was not able to distinguish quality and hence reward merchants and producers on that basis.[27] But the National Coffee Board enacted several coffee cleaning and grading regulations just two years after its establishment. The main purpose was to ensure that the processing and export of quality coffee were in accordance with consumers' preference.

According to the regulation, no coffee could be certified as "cleaned and graded" unless it contained less than 11.5 per cent moisture by weight and unless it was free of mould and

---

26. *Ibid.*, p. 15.
27. The Technical Agency, IEG, Marketing of Grains and Pulses in Ethiopia, *op. cit.*, p. 115.

other negative effects resulting from such defects. All cleaned and graded coffee was to be classified according to origin and grade. Origin referred to usage of the trade name such as Marrar, Sidamo, Jimma, ete. The bags were marked with one of these names.[28] There were eight grades which were determined by the number of defects found in a 300 gram sample. Grades 1 to 5 were described as Usual Good Quality and Ethiopia exported such coffee. Such measures were designed not only to earn sufficient foreign exchange, but also to be able to compete better with ever-improving quality of coffee.

TABLE VI.1

Different Grades of Coffee with Defects

Grade	Defects
Grade 1	0 to 3 Defects
Grdae 2	4 to 12 Defects
Grade 3	13 to 25 Defects
Grade 4	26 to 45 Defects
Grade 5	46 to 100 Defects
Grade 6	101 to 153 Defects
Grade 7	154 to 340 Defects
Grade 8	over 340 Defects

To ensure that these regulations were strictly followed, the National Coffee Board had stationed about sixty field officers in the major coffee producing and marketing centres. The main duty of these officers was to go to interior parts of the country and advise farmers to cultivate, prune, pick, wash and dry their coffee in the proper manner. It was also, a part of their duty to see that the coffee was properly dried, packed and sent straight to the Inspection Centre in Addis Ababa for final checking. There was also one centre at Dire Dawa. The National Coffee Board was also encouraging farmers to wash their coffee by acquiring pulper and wires for them for drying coffee at cost price. Farmers could even purchase these items on credit and pay for them over a liberal three year period. Due

28. Ministry of Information, Agriculture in Ethiopia, Book II, *op. cit.*, p. 13.

to the advice and help of the field officers to the farmers and merchants, there existed more than 174 hulling and 104 pulping stations throughout the coffee growing areas.[29] The hullers and pulpers had been of considerable help in improving the quality of coffee in the country.

Price Determination

The factors which determined the prices of grains were different from those which determined the prices of coffee and pulses. Ethiopian grain prices were very largely determined by the interaction of marketable surplus supply and demand. The world price and the Ethiopian government's intervention had insignificant influence on the price of grains.[30] As mentioned earlier, there were two terminal markets—Addis Ababa and Asmara, and prices throughout the country generally related to the prices prevailing in these two markets. The most important merchants involved in setting the general price level were the large assemblers and wholesalers who had trade mostly in Addis Ababa. "Their collective expectation of supply-demand relationship, along with their ability to support the market, is the most pervasive price determining force."[31]

The prices of coffee and pulses prevailing in the two terminal markets were not determined by the interaction of supply and demand forces but they were related to the prices available in the world market for these products. The merchants of Ethiopia had to compete not only amongst themselves but also with merchants of other exporting countries. These commodities had been subject to a fairly recurring price cycle : a period of scarcity and sharp rising prices followed by a period of surpluses and low prices.[32] These cycles affected the internal prices of these commodities.

---

29. The National Coffee Board, The Story of the National Coffee Board of Ethiopia, *op. cit*, p. 14,
30. The Technical Agency, IEG, Marketing of Grains and Pulses in Ethiopia, *op. cit.*, p, 107.
31. *Ibid.*, p. 107.
32. FAO, Commodity Bulletin No. 33, The World Coffee Economy, Rome, 1961, p. 7,

Margins and Costs

Once the prices of coffee and pulses are determined, it is easy to derive the margin of marketing. The total gross margin of the entire marketing system is the difference between the price received by the producer and the price paid by the ultimate consumer. In the case of pulses and coffee, the price is taken for consideration at the port. If the value of inputs purchased from other sectors including transportation are deducted from the gross margin, the remaining amount is the contribution of marketing to the total output of the economy. The factors which determine the margin of the marketing system are as follows.[33]

(*a*) The distance between the location of producers and the ultimate consumers determines the cost of transportation incorporated in the value of the product.

(*b*) The location of the producers in relation to the market. If the producers are near the large markets, the margin of marketing will be less.

(*c*) Margins will be higher when the supply is large in relation to demand and lower in the reverse case.

(*d*) If there is heavy competition between buyers and sellers and there is perfect knowledge about market trends, the margin will be less.

(*e*) The longer the produce is stored by the marketing system, the higher will be the total marketing margin.

(*f*) The greater the transformation of the produce by the marketing system, the larger can be marketing margin expected to be.

The marketing system prevailing before the Revolution show d that the producers sold their products in large quantities in relation to their demand in most of the primary markets. They did not have much knowledge about the prices prevailing in other markets for their products. The cost of transportation was very high. All these factors were

33. The Technical Agency, IEG, Marketing of Grains and Pulses in Ethiopia, *op. cit.*, pp. 134-135.

responsible in making the margin as high as possible. The producers were paid only a small amount of their products.

The Stanford Research Institute collected data on the cost of a pulses export transaction.[34] Table VI.2 shows that the margin of marketing was as high as 72 per cent. A higher margin would directly reduce the share received by the producers. In this case producer had received only 28 per cent. The distribution of the margin among different marketing agencies shows that transportation and handling absorbed about 39 per cent (22 per cent in Ethiopia and 17 per cent in shipping). The merchants got about 8 per cent. Cleaning, weight losses and packing accounted for about 12 per cent and direct taxes were about 6 per cent of the London price.

TABLE VI.2

Typical Costs of a Pulse Export Transaction

		Percent
Selling price in London—		100.0
Less Sea freight and insurance (Mitsiwa-London)	15.8	
London Agent's commision	1.0	
Supervision fees for weighting	0.4	
Loss-in-weight in Transit	1.2	
Interest on capital	2.0	
Bank charges (exchange)	0.8	
Insurance premium	1.0	
Chamber of Commerce Stamp	0.3	
Export duties on f.o.b. value	2.0	
Shipping charges at Mitsiwa (loading)	0.8	
Transportation (Asmara-Mitsiwa)		
— Warehouse to railway station, weighting, handling and loading	0.8	
— Railway freight	4.2	
— Unloading at Mitsiwa	0.3	
Asmara buying broker's brokerage	0.3	
Exporter's margin	3.0	
Exporter's total costs and margin		34.0
		66.0

34. *Ibid.*, p. 134.

Asmara price—		
Less cleaning expenses	1.7	
Handling-weight, load and pack	1.7	
Asmara selling broker's brokerage	0.3	
Loss of weight in cleaning	4.2	
Wholesaler's margin	1.7	
Wholesaler's margin and costs		9.6
		56.4
Cost on arrival in Asmara—		
Less Transportation from Gondar	13.3	
Municipality tax in Gondar	3.3	
Bags (10)	4.7	
Handling	0.3	
Assembler's margin	1.7	
First Assembler's costs and margin		23.3
		33.1
Gondar price		
Less Transportation from primary market	3.3	
Assembler's margin	1.7	
Second Assembler's costs and margin		5.0
Primary market price near Gondar		28.1

Source : The Technical Agency, IEG, Marketing of Grains and Pulses in Ethiopia, Stanford Research Institute, California, 1969 p 143.

The National Coffee Board also prepared a table indicating the cost of one kilo green coffee. Table VI.3 shows that the margin of profit to be shared between producers, middlemen, and exoprt merchants amounted to Eth. $ 1.6 per kilo of green coffee-average price Eth. $ 2.2 per kilo on the foreign market minus Eth. $ 1.6 the price prevailing in Addis Ababa market.[35] The further break up of the margin is not available, but it is clear that due to the functioning of co-operative societies in the producing areas and merchant committees, the producers might have received better shares as compared to producers of other commodities. The most important expenses were also transportation and handling which totalled about 16 per cent.

35. International Bank for Reconstruction and Development/IDA, The Economy of Ethiopia, Vol. II-Agriculture, *op. cit.*, p. 19.

TABLE VI.3
Cost of One Kilo of Green Coffee

(In Eth. $)

Items	Plantation	Semi-Plantation
Labour	0.188	0.167
Land tax and rent	0.026	0.010
Equipment	0.004	0.007
Haulage	0.020	0.020
Transport to local market	0.018	0.015
Transport to Addis Ababa	0.050	0.054
Total	0.306	0.273
Cleaning, hand picking and handling	0.020	0.024
Transport to Port and charges of f.o.b.	0.020	0.020
Total	0.346	0.317
Total Taxes	0.283	0.283
Total	5.629	0.600
Average of f.o.b. price	2.270	2.270

Source : The National Coffee Board, quoted from IBRD/IDA—The Economy of Ethiopia (in five volumes) Vol. II—Agriculture, 1967, p. 19.

Conclusion

These imperfections partly help to explain the failure of production of grains and pulses and, in fact, other commodities to keep pace with the demand before the Revolution. In the case of coffee, the production had increased but the quality could not be improved to the maximum. "Ethiopian producers have shown that they respond to price incentives, but have generally not received from the marketing system of price attractive enough"[36]

The Ethiopian government had taken some steps to improve the marketing system by establishing the Ethiopian

36. The Technical Agency, IEG, Marketing of Grains and Pulses in Ethiopia, *op. cit.*, p. 11.

Grain Corporation, the National Coffee Board and co-operative societies. The main purpose was "to encourage the production of agricultural products throughout our empire by stabilising the market, improving the quality of agricultural products, and by exporting agricultural products."[37] The Corporation was empowered to purchase, grade, process and transport agricultural products including grains, pulses and oilseeds. With a net-work of storage facilities, it was thought that the Corporation would be able to purchase all the offered marketable surpluses and intervene in the market wherever necessary. Similarly, the co-operative societies could easily obtain credit from the banks and thus avoid the selling of products at the time of the post harvest period.

But such planning is seldom matched by its implementation. The National Coffee Board could not start its function properly regarding purchasing, procuring, storing and marketing of coffee. The Corporation was in fact established in response to the very major decline in Ethiopian grain export with the intention to encourage production and export and it was in the initial stage in achieving its objectives. But the operations of the Corporation had not reached a point "where they are beginning to have a major impact on the marketing system."[38] On the other hand, the Corporation suffered from the shortcomings related to shortage of working capital, the lack of a government price policy and insufficient market information, and it was unable to generate enough profit to cover its administrative overhead costs.[39] The co-operative societies though functioning in some of the important producing centres, were also in their early years. They were also facing problems like shortage of funds and trained personnel. But these societies, if fully developed, could certainly have a significant influence on the marketing system.

37. Notice No. 267, of 1960.

38 The Technical Agency, IEG, Marketing of Grains and Pulses in Ethiopia, *op. cit.*, p. 105.

39. John Holmberg, Grain Marketing and Land Reform in Ethiopia, *op. cit.*, p. 10.

CHAPTER SEVEN

Community Development Programme

IN UNDERDEVELOPED countries the efforts of the government are recognised to be an indispensable force for development purposes. But in agriculture, with its large number of small and scattered production units, the control exercised by the government is naturally weaker than in industry. It is therefore not easy to relate agricultural development to a planned national development programme as a whole. This in part explains the interest of the government in other agencies like community development. If a well conceived community development programme is part of a national development programme, it should be able to provide answers to most of the basic problems of the economy. "Community development as a process is designed to stimulate and promote conditions for social, cultural and economic progress."[1] It can help, for example, to increase production

1. United Nations Series on Community Development, Aspects of National Community Development Programmes in Asia, New York, 1959, p. 2.

of agriculture if there is enough emphasis on agricultural extension and on economic and social incentives to which farmers respond. As such it could, given the right conditions, contribute to growth.

The term community development has been widely used but has meant a variety of things in different countries. The approach to and emphasis on the different facets of the programme vary in different countries. The United Nations and specialised agencies define it as follows : "The term 'community development' has come into international usage to connote the processes by which the efforts of the people themselves are united with those of governmental authorities to improve the economic, social and cultural conditions of communities, to integrate these communities into the life of the nation, and to enable them to contribute fully to national progress......This complex process is then made up of two essential elements : the participation of the people themselves in efforts to improve their level of living with as much reliance as possible on their initiative ; and the provision of technical and other services in ways which encourage initiative, self-help and mutual help and make them more effective. It is expressed in programmes designed to achieve a wide variety of specific improvement."[2]

There are five important aspects of this definition. Firstly, the local people and the Government should contribute towards the development of communities. The people on their part may contribute labour, management, cash and sometimes land. Government authorities may contribute technical assistance, equipment, grants-in aid and loans. This definition excludes local self-help and other activities without the assistance of Government. Secondly, the concept of the community development programmes is broad in its aims and methods. They are long-term programmes and have multiple objectives. "It is a planned, co-operative

2. United Nations Community Development and Economic Development, Part 1—A Study of the Contribution of Rural Community Development Programmes to National Economic Development in Asia and Far East. Bangkok, 1960, p. 1.

effort to advance the overall well-being of the community."[3] Earlier programmes sometimes put much emphasis on mass education and exclude other important aspects. Now, in fact, fundamental education and agriculture and, where possible, other extension and co-operative societies have been drawn in as necessary components of the community development complex.

Thirdly, the people should not only participate in the efforts made but should use their initiative as much as possible. "In the literature of community development, this concept has often been linked with the concept of the people's 'felt needs' which, it is said, should determine the lines of action."[4] Fourthly, attempts to secure action through pressure and compulsion are out of this definition. It is a programme based on the voluntary association of the communities. Lastly, this programme should have adequate support by way of national policy or it should be integrated with the overall national programme of development.

The Asian Programmes

During the last two decades, community development programmes have been implemented in most countries in Asia. They differ in content because the genesis of any programme is conditioned by the purposes with which it is started and the setting within which it operates. The most significant community development programme is India's. Its purpose is not only to expand the rural economy but also to smooth the course of social change. According to the First Five Year Plan, "Community development is the method and rural extension the agency through which the Five Year Plan seeks to initiate the process of transformation of the social and economic life of the villages."[5] There are blocks composed of

---

3. United Nations, Handbook for Community Organisation and Development Leaders (with reference to underdeveloped areas), New York. 1953 (First Draft), p. 1,
4. United Nations, Community Development and Economic Development, *op. cit.*, p. 3.
5. Government of India, Planning Commission, First Five-Year Plan, New Delhi, 1952, p. 102.

extension workers for agricultural purposes and village level workers for rural welfare. There are also people's organisation in the villages. As such the double emphasis on bringing together technical specialist within the reach of the villagers, and on the collective initiative and participation of villagers clearly bears a close relation to the United Nations' definition. In Pakistan, the community development programme is known as the Village Agricultural and Industrial Development Programme. This programme, like India's, lays emphasis on expanded production as an essential part of multiple objectives. Village level workers are given an important place in it.

The programme of community development in the Republic of Vietnam differs from that of India and Pakistan. When this programme was initiated, there were about 2,000 civil action agents working in the villages in mobile teams to provide political security and civil education. Later on, most of the agents were assigned community development work. The most important activities under this programme are construction and reconstruction of roads, schools and other infrastructure items. Now efforts are being made to establish "Farmers' Association" for this purpose.[6]

The community development programme in Ceylon, one of the oldest programme, is based on local leaders rather than on a corps of paid village-level government workers. There are rural development societies whose main functions are to give their members a new form of neighbourhood life. Women's societies are also operating in these villages along with rural development societies.

A different pattern has been evolved in Taiwan. The programme is strongly oriented towards promoting economic activities among the rural communities on a business basis, the main agency being "Farmers' Associations". Their functions include marketing, warehousing, processing of farm products, receiving deposits from and extending credit to members. Though operated on a democratic and co-operative basis, the societies are organised mainly on Government

6. United Nations, Community Development and Economie Development, *op. cit.*, p. 6.

initiatives and work under close government supervision and guidance.[7] Other important countries where community development programmes have been started, are the Philippines, Burma, Nepal, Thailand, Cambodia and Malaysia. All these programmes differ in content, organisation and purpose. It is difficult to make generalisations. In some programmes, economic development is less important than other aspects. The degree of direct attention for raising agricultural output also varies.

Ethiopian Framework

The above analysis shows that the conceptual framework and methodology of a community development programme depends upon the circumstances in which a country finds itself at a given time. When the community development programme was implemented in Ethiopia towards the end of the First Plan, it was felt necessary to limit it to improving the lot of the nomadic population and people in special areas, those areas which are not conducive to special progress due to natural factors.[8] The programme was to include the construction of wells as well as advice and technical aid for farming. Improvement of communications was also felt necessary in these areas. On the other hand, the needs of consumption could hardly be overlooked and provision for impovement of housing, nutrition and hygiene as well as construction of schools had to be made. Moreover, for the mobilisation of support of rural areas, their participation was essential to its success which could be achieved by setting up development committees in some areas to guide the programme. The First Five Year Plan defined the community development programme thus : "Community Development Progrmmes should play an important role in the overall development of the country and particularly in rural development. The main purpose of these programmes is to create on the local level the conditions required for accelerated economic, social and cultural progress. The

7. *Ibid.*, p. 6.
8. Ministry of National Community Development and Social Affairs, IEG, Proposals for Third Five Year Plan, p. 12.

experience of many underdeveloped countries suggests that full results in the economic sphere, public health, education and standard of living can be achieved only if co-ordinated and overall efforts are undertaken in various local units such as districts, communities, etc. for rural development."[9] The Plan emphasised the following aspects of the programme.

1. Water sources should be harnessed, water wells and channels dug, public fountains constructed, and other projects carried out for an improved water supply. Water troughs should be provided for cattle.
2. In the villages, streets should be repaired, ditches dug to carry away rain and pits sunk ; trees should be planted along the streets and around villages to provide protective forest belts, firewood etc. ;
3. To ameliorate conditions in selling agricultural products, common markets should be reconstructed, covered spaces and stores should be erected for exhibiting products ; markets should be provided with scales ; organised services should be established for maintaining cleanliness ; cattle markets should be separated from other markets ;
4. Townships should be electrified wherever possibilities exist ;
5. A standard type of school building should be erected for elementary education ;
6. Public health centres and infirmaries should be set up in some centres ;
7. Smaller buildings should be erected for the use of local authorities ;
8. Small economic projects should be established to serve the local population, such as water and mills stores for agricultural products, particularly in purchasing centres, small oil presses, slaughter-houses, nurseries for fruit-trees and afforestation, livestock breeding stations, veterinary stations, etc.

---

9. Imperial Ethiopian Government, Five-Year Development Plan 1957-61, *op. cit.*, p. 120.

9. Local roads linking villages with each other, with cities, and with main lines of communication should be built and repaired.

Utilisation of Resources

This conceptual framework and methodology of the programme was refined in the Second Plan. In view of the small domestic savings of 5 per cent of the national income and also of the limited resources of the country for development, it was realised that unless the local human and material resources were utilised in the cause of development, there would be no prospects of bringing about a rapid change in the way of life and work of the Ethiopian people. "Mobilisation of local human and material resources spurred by matching grants-in-aid from the Government, therefore, became one of the basic principles of our community development programme".[10] In the field of agriculture the programme laid particular emphasis on propagating new and improved techniques of production which could transform the subsistence sector of the economy into a modern one.

The programme also encouraged formation of co-operative farms in the countryside and included schemes of settlements in special areas in order to culturally integrate the nomadic sections of the people with the rest of the population. It continued to include basic education for villagers and added child-care, nutrition, personal and social hygiene and improved housing. Emphasis was put on those social services which could directly improve economic activities. In other words, the programme was directed to pursuing the following goals and objectives during the Second Plan period.[11]

(*a*) To mobilise local resources, both natural and human, with least reliance on the government for assistance in order to accelerate the socio-economic development for the community cencerned ;

---

10. Ministry of National Community Development and Social Affairs IEG, Proposal for Third Five-Year Plan, *op. cit.*, p, 12.
11. Imperial Ethiopian Government, Second Five Year Development Plan, 1963-67, *op. cit.*, pp. 304-05.

(*b*) To make the people of the community realise the benefits of substituting new methods of production for the traditional ones, and thus to pave the way for the gradual transformation of the purely subsistence rural economy into a more monetised one, capable of contributing its potential share to the economic growth of the country ;

(*c*) To encourage the creation of co-operative societies in order to provide a wide and sound basis for overall development ;

(*d*) To provide basic education to the villagers in the field of primary schooling, hygiene, nurition, child-care, housing and orgnisation to satisfy collective as well as individual needs ;

(*e*) To encourage the people of the community to adopt the habit of self-help and self-reliance, and thereby change their psychological outlook of helplessness ;

(*f*) To encourage the widening of the people's demands for higher standards of living in material as well as other spheres of social life ;

(*g*) To make the rural community conscious of the importance of common aims and co-operative efforts;

(*h*) To increase the socio-economic development of the community so as to create favourable conditions for undertaking big projects and implementing them in these regions, and to ensure the most rational use of the resources provided by the Government ;

(*i*) To promote cultural integration of the diverse communities.

The Plan added that "the achievement to be aimed in the community development programme must be supplementary to the development projects initiated by the Government in improving the living conditions of the people."[12] The community development programme will have to create favourable conditions in which resources (local, human and material) are effectively utilised for the country's socio-economic advancement.

12. *Ibid.*, p. 305.

"Economic and social changes are inter-related, yet if the possibilities for achieving quick economic results exist, the emphasis on the services to the rural community that will directly contribute to the improvement of economic activities may be decisive in setting in motion an accelerated socio-economic development in all its important aspects. The fact that the agriculture sector represents a big potential source of raising the national income in the years to come calls for special attention to be devoted to the community development programmes in agriculture, with the aim of laying foundations for a speedy transformation of the traditional subsistence agriculture into the modern sector, without which the long-term sustained growth of the economy would not be possible. This direction of the general policy is also suggested by the substantial rise in the quantities of agricultural output, with an accompanying increases of the income of the rural people that can be expected through the introduction of the comparatively simple new techniques and methods of farming."[13]

Rural Transformation

During the Third Plan which started in 1968 it was thought necessary to develop the programme into an integrated total approach to rural transformation based on the improvement of individual and collective economic, social and cultural skills of the population in the countryside.[14] Maximisation of agricultural production and rural transformation were considered as important as industrialisation in achieving self-sustaining economic growth. For the achievement of this objective, some conditions had to be fulfilled. These were : Improvement in the rural skills of agriculture, quick and effective diffusion of improved practices in agriculture and allied activities ; provision of adequate grants-in-aid and closely defined schemes for different areas, including special areas. The Third Plan says :

"Community development is a programme of aided self-help for the improvement of individual and collective skills

---

13. *Ibid.*, pp. 305-306.
14. Ministry of National Community Development and Social Affairs, IEG, Proposals for Third Five Year Plan, *op. cit.*, p. 13.

of the rural population. The programme, to succeed and to have real benefit, must be developed into an integrated total approach to rural transformation, based on local resources and initiative.

The programme has therefore to be planned and implemented by the rural communities themselves, the Government supplying only such technical and financial help as is indispensable and essential.[15]

For the purposes of implementation, the programme was divided into different categories. In the first category came agriculture and allied activities. For the development of agriculture, the programme had to concentrate on the following : (a) Demonstration and popularisation of improved agricultural implements including sprayers and dusters ; (b) layout of ocular demonstration ; (c) purchase, maintenance and repairs of equipment for demonstrations to be carried out in farmers' fields ; (d) encouragement to farmers' clubs and promotion of progressive farmers' training in the fieled of agriculture ; and (e) provision of fruit-growing and vegetable cultivation.[16]

Organisational Approach

In the implementation of the community development programme the method and approach were clearly specified. The multi-purpose community development workers, known as village level workers, were the main agent of implementation of the programme in view of the shortage of trained specialised personnel. These workers were trained at the Training and Demonstration Centre. The co-operation and participation of the people and the establishment of village development councils were also considered for the success of the programme. Adoption of such techniques as demonstrations, group approach, direct contacts and distribution of literature, would create a sense of self-help and self-reliance in the minds of the people.

15. Imperial Ethiopian Government, Third Five Year Development Plan, 1968-73, *op. cit.*, p. 322.
16. Ministry of National Community Development and Social Affairs, IEG, Proposals for Third Five Year Plan, *op. clt.*; p. 21.

The Board of National Community Development consisting of the Ministers of Agriculture, Health, Education, Interior and Public works was the highest body for the co-ordination of the community development programmes. The Minister of Community Development acted as Chairman of the Board. The inter-ministerial co-ordination committee consisted of executive officers appointed by their respective ministries among the members of the Board, assisted it in implementing its decisions. The Community Development Department of the Ministry of National Community Development and Social Affairs was concerned with field operations, adult literacy and education, aid programmes in special areas, etc.

The unit of operation of the programme was the district and the District Development Officer was the head of the programme to report direct to the Ministry. Considering that this centralised operation in the Head Office of the Ministry was not a satisfactory arrangement from the point of view of active, on-the-spot supervision and taking of quick decisions, a beginning had been made in the Third Plan to decentralise the administration of the programme by establishing four regional offices of the Community Development Department, each incharge of three or four provinces as a prelude to the setting up of Provincial Community Development Offices.

Initially, there were 10 village level workers under the supervision of the Community Develppment Officers per worda (district) project, each worker catering for about 1800-2000 of population. This was necessary when the programme was in an experimental stage. Later, it was decided to assign only five male and two female workers per worda on a uniform basis. This would enable a faster and more economical coverage of a larger area within the available financial resources.

At the district level, the worda Governor acted as the Chairmen of the District Development Committee. The District Development Officer acted as the Executive Secretary. The Chairman was responsible for the co-ordination of all the district development activities. A district usually covered 10 Atibas (Villages) or 20,000 population. In each Atiba there was an Atiba Council which was the smallest unit of the programme. The Atiba Shum (village leader) acted as its chairman. Members were elected by the villagers. The village level worker was secretary of the Council. The Council was responsible for co-ordinating the programme within the Atiba.

Finances of the Programme

In the programme of community development, the fullest mobilisation of available resources such as labour, material, and cash must be the underlying principle and government assistance should be looked for only when and where it is indispensable.[17] But at the beginning of the programme, Government assistance is very essential for expansion. Financial assistance from the Government implies that the selected district will be allocated a certain amount every year as working funds and a lump sum for the five-year period.

During the First Plan a total allocation of Eth. $25 million and a supplementary allocation of Eth. $ 8 million were provided. It was designed to cover about 50 to 100 out of 1955 sub-districts. At this rate, it was expected that the entire country could be covered over a period of 30 to 40 years. Regarding the contribution of local people, it was supposed that it would be almost equal to the value of the central fund. The total value of the local development projects was estimated at Eth. $ 33 to 66 million for the following five years. The Plan was to provid Eth. $ 25 million. The expected local contribution was to be about Eth. $ 25 million.[18] But unfortunately this proposal did not materialise due to financial and other difficulties. However, the Ministry of Community Development had manged to begin the programme with the small budget of Eth. $ 1.3 million a year. The contribution would have been more if a proper campaign had been launched.[19]

---

17. Imperial Ethiopian Government, Second Five Year Development Plan, 1963-67, *op. cit.*, p. 312.
18. Office of the Planning Board, IEG, Five Year Community Development Programme, Documentation No. DOG-55-COM-3, p. 8
19. Office of the Planning Board, IEG, National Community Development Scheme in the Second Five Year Plan, Documentation No, SS-COM-2, April, 1962, p. 3,

The total financial resources to be employed for the programme during the Second Five Year Plan amounted to Eth. $ 25.5 million, out of which about Eth. $ 14.7 million was recurrent expenditure, while Eth. $ 10.8 million was meant for investment purposes. The above sum was apportioned as follows :

TABLE VII.1

Recurring Expenditure and Investment During Second Five Year Plan

Item	ETH. $ Recurrent	Investment
Community Development	10,020,000	9,866,000
Training of C.D. Personnel	1,504,000	153,498
Social Welfare	2,625,000	675,000
Training for Social Welfare Programme	282,000	10,502
Labour Exchange	344,000	68,000

Source : Imperial Ethiopian Government Second Five Year Development Plan 1955-59 EC (1963-67), Addis Ababa, 1962, p. 312.

One of the most desirable aids that government could give at the initial stages of the Community Development Programme was the establishment of a network of model farms, such as those already set up by a Government agency at Awassa.[20] These projects formed part of the programme of agriculture but wherever possible they served to demonstrate advanced farming techniques to the villagers and train development workers and young agriculturists. There was provision to train some 900 agricultural technicians and 175 agricultural experts who should be college graduates in order to meet the demands of the programme during the Second Plan. But apart from the expenditure on the limited organisational development, the

20. Imperial Ethiopian Government, Second Five Year Development Plan, 1963-67, *op. cit.*, p. 312.

amount allocated for the Plan could not be spent. As against Eth. $ 25.5 million for a projected 40 to 50 additional centres, the actual allotment was only Eth. $ 8 million.[21] The quantum of grants-in-aid in the form of technical services and equipment in support of self-help schemes was relatively small, for after meeting the committee expenditure on the staff and other essential capital items, little was left for the development scheme. In the field of agriculture and allied services, irrigation canals and grain stores, well construction and co-operative societies were organised. In one year, 1967-68, the total cost of all such projects was over Eth. $ 270,000, of which over Eth. $ 255,000 was contributed in cash and kind by the rural communities.

During the Third Plan the total planned expenditure on community development was to be about Eth. $ 22 million of which Eth. $ 16.1 million in the form of recurrent items and Eth. $ 5.9 million in capital outlays.[22] The recurrent expenditure was composed mainly of the salaries and wages of the district development officers and their assistants as well as village level workers. The total expenditure was to be about Eth. $ 13.6 million during the five years and the remaining expenditure would be on items such as two training and demonstration centres. Of the Eth. $ 5.9 million capital expenditure, Eth. $ 2.9 million was for the construction of community development centre projects and Eth. $ 0.8 million for the expansion of the Training Centre at Awassa as well as for the construction of new training centres and the establishment of Youth Centres. In addition the government was to furnish grants-in-aid amounting to Eth. $ 2.2 million.

Progress Achieved During First and Second Plan

Since a sustained increase in total output per capita provides direct evidence that the economy is becoming more productive, the question that arises is how much increase in production has actually resulted from the community development programme. So the problems of measurement immediately

---

21. Ministry of National Community Development and Social Affairs, IEG, Proposals for Third Five Year Plan, *op. cit.*, p. 12.
22. Imperial Ethiopian Government, Third Five Year Development Plan 1968-73, *op. cit.*, p. 357.

appear. Since the data about different aspects affecting production and their break-up are not available, only some trends can be analysed. Further, for some kinds of output the increase cannot be measured directly. Thus, changes in the farm crops are sometimes roughly calculated on the basis of changes in area tilled or the extent to which improved seeds and fertilizers are used.[23]

This analysis clearly shows that the community development programme during the First Plan was not output-oriented. It concentrated on literacy and adult education. Some steps for improvement of the lot of the nomadic population were also taken. So the activities that were only complementary to production dominated the initial stages of the programme. In September, 1962 the first batch of multi-purpose trained extension workers was put on the field. During the Second Plan period, the programme laid particular emphasis on propagating new and improved techniques of production in the field of agriculture and allied activities. The main emphasis on the "maximisation of agricultural production and rural transformation" was introduced in the Third Plan.

In the First Plan period, 17 district (Worda) development projects were launched in Special Areas (nomadic and semi-nomadic) with staff deputed from other ministries. Community development projects were staffed by priests and community teachers. By the end of the Second Plan the total number of District Development Projects in operation was 32 covering a population of one million. In addition, a Demonstration Centre was also in operation. The total number of District Development Committees was however somewhat smaller. At Atiba level, about 140 Development Councils were organised. About 151 village level workers (30 of them women) were trained during this period. They were able to organise 65 women's voluntary organisations with a membership of 2,000. These organisations undertook programmes of literacy, child-care and other welfare activities. Nearly 70 Youth Clubs carried out the youth programme.[24]

---

23. Uinted Nations, Community Development and Economic Development, *op. cit.*, p. 16.
24. Ministry of National Community Development & Social Affairs IEG, Proposals for Third Five Year Plan, *op. cit.*, p. 16.

A. Agricul ture

In the field of agriculture, irrigation canals, wells and grain stores were constructed and farms such as Arba-Minch, Awassa and Botter organised. In some of the areas the Community Development Department had distributed improved seeds but the amount was very negligible. In one of the reports of a team to review the progress of the Gowane Project which was designed to serve about 25,000 semi-nomadic people spread over an area of 19 kms it had been observed that "it is surprising that over the five years of its existence the Centre itself looks a desolate and a deserted place...the area has been subjected to drought for two consecutive years, yet no official relief measures have been initiated as it is the usual practice in the case of rural areas affected similarly by famine conditions. The people who had settled near the Centre have migrated away from it, whether it was due to drought and disappearance of grass land, or suspicions about the nature of the project or its workers or threat to their lands or to their culture or just insecurity...The people still practise barter in the economy and often the unit of exchange is either butter...People have to walk long distance to Awash station or Errar or Dire Dawa and the weekly market in Kamase or Senbete is 70 kms."[25] Similarly, in other centres there was not much progress. During the whole period, the community development workers distributed only Eth. $ 6,000 worth of certified seeds and Eth. $ 5,000 worth of vegetable seeds to farmers in the various centres of the country.[26]

B. State Agricultural Farms

During the Second Plan a policy of setting up large-scale agricultural farms on commerical lines was adopted. Apart from exercising influence on the countryside in modernising the techniques of agricultural production, thus serving as the spearheads of an economic and social transformation of the

25. Ministry of National Community Development and Social Affairs Report of a Team Review of Progress of Community Development Activities in the Gowane Worda, Addis Ababa, 1967, p. 3-6.
26. Ministry of Planning and Development Community Development & Social Welfare, Documentation No. SS-COM-18, p. 36.

rural sector, these farms were expected to meet many other needs-checking the rural exodus by improving the employment opportunities of under-employed and seasonally unemployed rural manpower, helping to generate cash incomes, thus carrying on the task of monetisation of the subsistence sector a step further.[27] Two important multi-crop state agricultural farms were established in Awassa and Arba-Minch and the foundation laid of a cattle farm in Botter.

1. Arba-Minch Farm

Before the Ministry of National Community Development and Social Affairs took over the project in 1962, known as Arba-Minch Farm, it was in the hands of the Sabean Utility Corporation. In 1960, the Corporation started systematic experimentation with cotton in this region. The selection of the region was due to the availability of water which is indispensable for producing cotton. Initially, the total cultivated area was 4 hectares yielding 35 quintals per hectare. Next year about 11 hectares were brought under cultivation. The Corporation wanted to start large-scale farming, but failed due to some "technical difficulties" and ultimately the Ministry of National Community Development ventured to start a pilot farm and a settlement scheme.

The development of this region was done in stages. In the initial stage the main efforts were directed towards the clearing of land and construction of an irrigation system to cover 170 hectares of land. In 1968 the irrigation system covered 830 hectares. The Arba-Minch Farm was established to achieve the following goals.[28]

1. Better distribution of the working agricultural population in relations to the agricultural potential of the region ;

27. Ministry of National Community Development & Social Affairs, Proposals for Third Five Year Plan, *op. cit.* p. 32.
28. Department of Agro-Industrial Development, Ministry of National Community Development and Social Affairs, Production Plan of Arba-Minch Farm 1969-70, Economic Section, Addis Ababa, 1969, p. vi.

2. Development of economic relations between the various regions of the country ;
3. Exploiting the geographical assets of such regions so as to assure regional specialisation ; and
4. Developing adequate, diversified and balanced economic activities.

Another objective was to facilitate the settlement of potential farmers within the inhabited area. The settlers' units were established on the basis of 3.20 hectares each family. The farm was to provide the settlers with seeds, pesticides, machine services and adviser and to find a market for their produce. The farm was authorised to cover all its expenses from the income earned and the remainder was to be shared equally between itself and the settlers.

The total area for disposal under the Ministry was 1,750 hectares. The area under production in 1969-70 was 960 hectares of which 746 hectares was under first crops and 214 hectares under second. In addition, 164 hectares of the area under the first seasonal crops was for the Second crops. Thus the total cropping area was 1,124 hectares. Cotton was the single most important cash crop. Corn was grown mainly as a staple food crop for the settlers and beans for rotation purposes. Teff was produced as a slack crop in rotation. Cotton occupied 2/3rd of the area. In order to see the progress of the farm, two important aspects have been discussed here : (*a*) the payment that went to each settler after paying all expenses incurred on his behalf by the farm, and (*b*) the number of settlers that remained with the programme.[29]

Table VII.2 shows the yield of different crops on the Arba-Minch farm and on the farms maintained by the settlers. It will be clear that the yield rates on the Arba-Minch farm had been increasing though the rate was low. There was two important crops grown on this farm ; beans and teff. The yield rate of both the crops had increased but it was far less than the planned rate. On the settlers' farms, cotton and beans were produced. The yield rate of these crops was much

29. *Ibid.*, p. viii.

higher than the farm and was more or less equal to the planned rates.

TABLE VII.2

Yearly Produce and Planned Yields at Arba-Minch Farm

(In quintals per hectare)

Name of Crop.	Produced					Planned	
	1964-65	65-66	66-67	67-68	68-69	68-69	69-70
			FARM				
Beans	0.07	1.66	8.52	5.27	N.A.	10	10
Teff	4.60	9.33	5.09	7.06	N.A.	10	10
			SETTLERS				
Cotton	5.50	12.73	15.24	16.28	15.37	16	16
Beans	6.16	5.10	5.56	7.74	N.A.	10	10

Source : Department of Agro-Industrial Development, Ministry of National Community Development and Social Affairs, Production Plan of Arba-Minch Farm, 1969-71, p. 12.

Many factors had contributed to this progress : Insect control programmes, irrigation schemes, improved seeds and other technical facilities. Without a good insect control programme, high yields could not be expected from cotton plantations. So spraying of insecticides was begun with the introduction of cotton on the farms in 1964. Since the area was small, spraying was done manually by Knapsack Sprayer. But this required very strict supervision. The irrigation scheme for the farm was designed to cover a gross area of 1,252 hectares. The main reason for limiting the area was the estimated available water from the Culfo river and the introduction of two crops per year. However, the measurement of water in the last few years had opened up a favourable possibility. So the expansion of an area of about 500 hectares was also considered.

It can be claimed that the establishment of the Farm had contributed largely to the development of the region either directly or indirectly through its effect on yields and area and supply of improved seeds, adoption of improved techniques and irrigation facilities.[30] But various problems had arisen from time to time particularly in connection with the settlers. The Farm was to cover all expenses from its income and the remainder was to be shared equally between the Farm and the settlers. But the Farm charges, it was estimated, had been much higher compared to its output. The limited area under cultivation, coupled with a very low yield per hectare, was the main factor responsible for the high cost of input. In 1968-69, the total receipts from the sale of cotton were Eth. $ 84,480 from the settlers' farms and teff and corn worth Eth. $ 24,000 was produced which was used for home consumption. The total expenses incurred on settlers' farms were estimated at Eth. $ 55,485. After deducting the expenses, very little was left for distribution among the settlers and the Farm.[31] Moreover, there were other condittion which the settlers were supposed to fulfil. On account of all these factors, the number of settlers decreased from 65 to 44 during this period. The new regulation fixed prices for the Farm's reproduction materials delivered to the settlers as well as for produce taken over from the settlers. It was assumed that in future the Farm's charges would come down following the research being done on this subject. Efforts had also been made by the management of the Farm to normalise conditions.

2. Awassa Farm

The Awassa region was covered by various projects. It had an area of about 16,000 sq. km. The Community Development programme took on a more specific character with the establishment of an intensive farming centre at Awassa in 1962. It had in the past few years been the major factor in Awassa regional development planning (especially for the settlement of semi-nomadic population groups) and in the establishment

30. Addis Ababa Chamber of Commerce, Arba-Minch Agricultural Farm, Ethiopian Trade Journal, Vol, II, No. 4, June, 1969, p. 23.
31. Department of Agro-Industrial Development, Production Plan of Arba-Minch Farm, *op. cit.*, p. viii.

of the town of Awassa, which had become the provincial capital.

The farm's economy was based on two main products : maize and sisal. To secure a higher income an attempt had been made to grow pimento. Other crops were sunflower, colze, beans and teff. Some of the area was also under tobacco cultivation. In 1961-62, the total area of the farm was 4,000 hectares, of which 3,600 hectares were reclaimed and 3,200 were put under cultivation. There were 427 veteran soldiers' families settled there. In 1967-68 the total area cultivated was 4,250 hectares.

The yield of maize on the farm was about 35 quintals per hectare. Research experiments on larger plots had recorded yields of 60 quintals per hectare. On this basis, it was estimated that the total production of maize on the farm would be about 6,000 tons in 1970. It was also expected that about 300 hectares under sisal would be reclaimed. The cost of production of maize was competitive compared to that of the exporting countries. Sisal production had not achieved much increase. Because of the existing marketing conditions and a single stripping machine, the total production of sisal was only 600 tons. The overhead expenses of production were very high in the past with the result that there was a very small margin of profit.[32]

The farm was not functioning efficiently due to two main reasons. First, a part of the quipment and machinery bought at the beginning of the development of the farm did not meet the requirements of the crops. Because of hard conditions of utilisation and lack of experience dpersons, machinery and equipment were out of order quickly. Secondly, the farm did not possess facilities for drying, storing, sorting, handling and picking of harvest products. The financial statement of the farm as of March 26, 1968, showed a deficit of Eth. $ 338,442.[33]

---

32. Ministry of National Community Development and Social Affairs, Agricultural and Industrial Development of Awassa Area, Final Report-Action in Peasant Life, October 1968, p. 10.

33. *Ibid.*, p. 48.

Slow Evolution

The foregoing analysis shows that the programme had not been able to achieve what was set out in the Plans both for community centres and for commercial farms. Consequently, it did not have a decisive influence on the growth of agricultural production in the economy. The main reasons were :

First, the programme as set out in the Second Plan laid more emphasis on attitude formation than on the preparation and execution of socio-economic projects. This means that the emphasis of the Plan was more on the building up of a strong organisation than on the expansion of the community development programme. Secondly, the community development programme was mainly the result of Government initiative. This does not alter the fact that many communities normally carry out and did carry out a lot of community work. But the contribution of the rural population was not as much as was expected. Thirdly, a major obstacle in the way of intensification of the programme during the Second Plan was the non-availability of adequate funds. As already mentioned, against an original estimate of Eth. $ 25.5 million, the allotment was only Eth. $ 8 million. As a result, the amount of grants-in-aid in the form of technical services and equipment for self-help schemes was relatively small, and after meeting the expenditure on staff and other capital items little was left for the development of the programme. Moreover, the financial rules of the government which stood in the way of automatic reappropriation of unsued funds from one year to the next partially accounted for its slow progress. Fourthly, an inadequate awareness on the part of heads of the general administration in the Worda and the Governor-General prevented the programme from gaining momentum in many places.[34] Lastly, the slow evolution of inter-ministerial co-ordination for the programme, also prevented it from having any deep impact in many areas. A National Board of Community Development was legally created and the effective use of this organisational set up was essential for the success of the programme as planned. But the

34. Ministry of National Community Development and Social Affairs, IEG, Proposals for Third Five Year Plan, *op. cit.*, p. 16.

National Board had been totally inactive throughout the planned period.

Proposals for Third Plan

The community centres together with the establishment of commercial farms, represented the first stage towards a policy of regional development. Naturally, the first task during the Third Plan was to consolidate, strengthen and put on a proper footing the existing programme and overcome the limitations imposed by the non-availability of resources and limited co-operation from other ministries. Alongside it was proposed that an honest effort to establish 67 new projects would be made, covering 100 out of 448 centres and making the benefits of community development available to about a fifth of the total population of Ethiopia.[35] To meet the need of qualified personnel, especially community development workers, the Training and Demonstration Centre at Awassa was expanded in the beginning of the Third Plan and it was expected that it would double its present output in the third year of the Plan. A Second Training and Demonstration Centre with a capacity of 160 was to be established during the second year of the Plan.

The dynamic role of the large agricultural state farms in relation to the employment generation, skill formation and propagation of new farming techniques on the one hand and monetisation, supply of raw materials to industry, export promotion and import substitution, on the other, pointed to the need for increasing their number and scale of operation in the Third Plan. "A preliminary study pointed out 23 possible areas for agro-industrial development covering practically all the 14 provinces."[36] But the immediate task during the Third Plan was to expand the Arba-Minch agricultural farm. This would also involve development of regions as a whole by providing extension services for farmers in the neighbourhood in respect of distribution of selected seeds, renting

---

35. Ministry of National Community Development and Social Affairs, IEG, Proposals for Third Five Year Plan, *op. cit.*, p. 10.
36. Ministry of National Community Development and Social Affairs, IEG, Proposals for Third Five Year Plan, *op. cit.*, p. 35.

of tractors, and buying their products at market prices. It would also help to maintain 44 settled families. The area under cultivation directly under the farm was expected to go up to 1,700 hectares by the end of the Third Plan. The Government also approved a programme to establish a processing complex at Awassa initially based on supplies from the farm. Another important programme was the planning and development of the Awassa-Blatte region. Factors like fertility of soil, availability of labour and the experience so far gained helped to undertake a preliminary study of the region. The study recommended that the water of the Awassa Lake should be carried through a tunnel of about 18 kms to Blatte River. It was estimated that about 4,000 hectares of land around the lake could be used for cultivation.

C. Package Programmes

During the Third Plan, the Government had accepted another approach, called 'package programme' of concentrated activities in relatively small, sharply defined regions for agricultural development, particularly among small peasant farmers. The package concept implies the co-ordination of a number of related activities such as provision of transport, marketing services, credit, as well as a related set of essential agricultural actions, such as distribution of better seeds, dissemination of information on improved agricultural practices, promotion of fertilisers etc. Such programmes do not have to cover all the suggested range of activities. They can be modest involving almost all spheres of Government and even private activities. The justification for such a policy of concentrating Government funds and activities in relatively small areas arises from other consideration as well. When skills and funds are scarce, a policy of concentration is desirable simply because there are not enough resources to cover the entire area of the country. So with substantial help from the Swedish International Development Agency and Ethiopian counterpart funds, the Government had established the Chilalo Agricultural Development Unit as a first step of what was planned to be a series of package programmes. The project in Arussi Province was a ten-year scheme to raise the level of peasant agriculture.

The CADU (Chilalo Agricultural Development Unit) programme was started in September 1967 based on a written agreement between the Imperial Ethiopian Government and the Government of Sweden. The purpose of the package programme was to establish within a geographically limited area suitable methods which, when applied in an integrated manner, would improve the ability of the local population to participate in and be responsible for the development of agriculture in the area. The Government, during the Third Plan period, was supposed to identify at least three additional package project areas and begin operations. The important areas under consideration were near Debre Zeit, Tigre, the land around Lake Tana, the northern area of Bale, Central Gojam. The SRI in its study on 'An Agricultural Credit Program for Ethiopia' also suggested other areas for the initiation of the package programme in Ethiopia.

The Military Government has abolished the community development programme as such. But it has realised the importance of state agricultural farms in transforming agricultural situation. These farms have been given a dominant position in the Agrarian Reform programme. The important objective of state farms is to speed up the transformation of agriculture by introducing improved techniques, supplying better seeds and setting examples for others. These farms are to produce crops not only for export purposes but also to meet the foodstuff needs of the population. The performance of these state farms has been discussed in Chapter VIII.

PART III

AGRICULTURE SINCE THE REVOLUTION

tration by the various trade unions. Students in the streets and teachers off the classes also showed signs of discontent in the country. In the countryside, the agricultural workers and landless tenants not only staged demonstrations but there were reports of their having burnt crops of the landlords. This was the time when the country suffered the highest death toll from the worst drought. The state-controlled information agency said that three to four million people had suffered from the "worst drought situation."

Simultaneously, serious allegations of corruption against the former Emperor were also levelled. He was accused of making money from the capital's bus company. The most serious of the charges was that the Emperor had sent abroad tons of gold to be hoarded in foreign banks. Estimates for this range up to 15 billion US dollars. Political unrest, corrupt administration, natural disaster and anti-feudal peasant riots were the main factors which eroded the long rule of the Emperor.

In the first few months following the overthrow of the Emperor, emphasis was laid on "Ethiopian Socialism." But from the middle of 1976, the authorities put forward the thesis of accepting scientific socialism based on the principles of Marxism-Leninism.[2] It requires profound socio-economic changes. Land, the principal source of the nation's natural resources, was nationalised and distributed among the peasants and landless people. All types of old landlord-tenant relationships were abolished. At the same time urban land and profitable apartment houses were nationalised. Key industries and leading enterprises were transferred to the state. Private bank and insurance companies were nationalised and old taxation system abolished. A progressive monetary reform was carried out, and all currency operations were put under government control. Private enterprise in industry and commerce was strictly limited and direct government influence on foreign and domestic trade became stronger.[3]

---

2. Georgi Galperin, Ethiopia : Population, Resources Economy, *op. cit.*, p. 18.
3. *Ibid.*, p. 18,

CHAPTER EIGHT

Agrarian Reform

IN SEPTEMBER 1974 the armed forces in Ethiopia dethroned Emperor Haile Selassie and appointed their own chief. The downfall of the Emperor marked a bloodless revolution as the armed forces eroded his power in a revolt. The Armed Forces Committee was formed which dissolved the bicameral Parliament and proclaimed a Provisional Military Government to run the nation.[1] These changes in Ethiopia had been "long overdue." For many years the people were disappointed by the lack of democratic representation within the Emperor's nation. Although he himself had a reputation as a man who fought for the national right of his people against the Fascist regime of Mussolini, his rein had hardly strengthened the cause of representative government.

Under popular pressure, he had agreed to a constitutional conference which was scheduled to report immediately about the political set-up. But the army, the only organisation capable of standing up to the Emperor, insisted on more changes. The demands of the army were quickly followed by demons-

1. For detailed information about Party Formation and Popular Participation, see Appendix I.

In 1979, Ethiopia started to implement the National Revolutionary Economic and Cultural Development Campaign considering it as the first step in a programme of peaceful construction and a transitional stage in laying the foundations of a socialist economy. In 1984, National Central Planning Committee has been established to make best use of country's human, financial and material resources under short and long term planning.

Land Reform

The Military Government also realised the importance of land reform as a necessary condition for Ethiopian modernisation. A person's status, and standard of living is determined by his relations to land and therefore it is essential to fundamentally change the existing land relations so that the peasants may be liberated from old-age feudal oppression and injustice. In order to lay the basis for equality, freedom and fraternity, and also to increase agricultural output and to make the tiller the owner of the fruits of his labour, it is necessary to release the productive forces of the rural economy by liquidating the feudal system.[4]

The Agrarian Reform Proclamation of March 1975[5] is therefore a historical act in the life of Ethiopian people. Under the Proclamation all rural lands were declared collective property of the Ethiopian people. Private ownership of rural land was abolished. All land was to be distributed without compensation to former owners among those who tilled it. Old landlord-tenant relationships were abolished and discrimination against women in land use rights was prohibited. The Proclamation also banned sale and purchase, succession, mortgage and renting of land. Privately owned large scale commerical farms were passed into the hands of state and co-operatives and were partially distributed among peasants. The size of one plot was not to exceed 10 hectares. All farmer tenants and farm labourers received the right to secure for themselves the land plots which they tilled when the Proclama-

4. Paul Brietzke, Land Reform in Revolutionary Ethiopia, The Journal of Modern African Studies, Vol. 14, No. 4, December, 1976, p. 637.
5. See Appendix II.

tion was promulgated. Tenants could retain for themselves at a reasonable compensation implements and a pair of oxen belonging to the former landlords. No person was allowed to use hired labour to cultivate his holdings. But this was not applicable in the case of women with no other adequate means of livelihood or where the holder is sick or old.

It is not enough just to declare the land public property and transfer it to the peasants. The peasants must be organised to carry out the provisions of the Proclamation. Peasants' Associations have been established. A peasants with less than 10 hectares of land is a member of this Association. Each Peasants' Association includes an average of 250-270 families, and each one is incharge of at least 800 hectares. District Peasants' Association are organised on the basis of local Associations. They are designed to assist and control land distribution; guide and co-ordinate local Associations' activities. Regional Associations are set up in sub-provinces and provinces to co-ordinate the activities of local and district organisations and execute government laws and orders. In 1978 an All-Ethiopian Peasants' Association was established. Its main task is to fundamentally transform agriculture, organise cultural and political work and also to encourage the co-operative movement.

The Peasants' Associations have the following basic functions in implementing land reforms : to distribute land among local peasants as much as possible equally ; to follow land use directives to be issued by the Government ; to administer and conserve any public property within the area especially the soil, water and forest ; to establish judicial tribunal to hear land disputes arising within the areas ; to establish marketing and credit co-operatives and other associations which would help the farmers ; to build with the co-operation of the government schools, clinics and similar institutions necessary for the areas ; and to undertake villagisation programme, etc. This shows that the Peasants' Associations have to achieve two important objectives, i.e. to control peasants politically and adopt certain patterns of land use, cultivation, harvesting and marketing. Now these associations have actually become self-governing bodies in the villages. In 1983, there were 19,600 Peasants' Associations with 5 million households.[6]

Co operatives and State Farms

Ever since the outbreak of the Revolution, the Government has been doing its level best towards the development of co-operatives. The Proclamation of December, 1975 states that no less than three and no more than ten Peasants' Associations may form an Agricultural Service Co-operative. The main objectives of Service Co-operatives are to buy the produce of members at a fair price, supply improved agricultural implements, make loans available at reasonable rates of interest, disseminate co-operative education in socialist ideology and supply consumer goods to members. Such co-operatives have been established by many members of the Peasants' Association. In 1983, there were 3,700 Agricultural Service Co-operatives with a total membership of 4.6 million heads of households. The total capital of all these Service Co-operatives was 92 million Birr. It had increased by more than 100 per cent compared to the 1981 figures.[7]

The Revolution has of course placed lands in the hands of peasants but they have very little draught animals and other farm equipments. The size of the plot is also not very large. In order to help the peasants and also to lay the basis for the building up of socialism, the Proclamation of June 1979 formulated the guiding principles for the establishment of Agricultural Producers' Co-operatives. These principles are to free small-holding farmers from the poverty and to make them supporters of socialism, to get them voluntarily and on their own to move from small, fragmented and individual ownership, stage by stage, to higher forms of co-operative ownership.[8] The initial stage called Malba Co-operative permits only partial amalgamation and joint use of farming tools, draught

6. Mengistu Woube, Problems of Land Reform Implementation in Rural Ethiopia, A Case Study of Dejen and Wolmera Districts, Geografiska Regionstudier, Nr. 16 Uppsala, 1986, p. 52.
7. *Ibid.*, p. 53.
8. Mulugetta Beyene, Strengthening Co-operative Movement, The Ethiopian Herald, January 29, 1982.

animals, etc. This stage allows members to retain upto 2,000 sq. metres of land for private use and turn the rest of their holdings to their co-operatives alongwith their domestic animals against the payment for compensation for the animals.

The second stage called Wolba Co-operatives socialises all tools and means of production. Cultivation of large tracts of land requires introduction of modern farming technique, organisation of common services and creation of permanent funds. The peasants can retain only 1,000 sq. metres of land for their personal use and no rent is received on the farm implements placed for joint use. The third stage called Woland will witness a complete transition to the socialist principles with the establisment of commune or Woland. The payment will be made "from each according to his ability to each according to his work". The size of this type of co-operative is envisaged to be high consisting of 2,500 members as against 500 members in the simple forms of societies. The private plots are abolished. But the available information shows that no Woland has yet been established in Ethiopia.

As far as the structural relationship is concerned, all organisations are linked to the various government institutions. All guidelines and plans come from either PMAC, the Worker's Party of Ethiopia or the National Committee for Central Planning and reach the people through various ministries and departments and organisations such as All Ethiopain Peasant's Association. The structural relationship between various organisations from the national level down to the village level is indicated in Diagram I.

In setting up producers' co-operatives, the major step is to ensure that the peasants join co-operatives willingly. They, in fact, are persuaded by the benefits from the co-operatives. In this regard the efforts made by political cadres are quite considerable. The peasants are given chance to determine the time when they would like to become members. Coercive measures are not supported neither from the political nor from the economic view point because such measures damage the very purpose of the movement. Further, if farmers are forced into membership, not only will their participation be low thus, reducing the effectiveness of the co-operatives, but such socie-

Diagram I : **Structural Relationship Between Different Organisations**

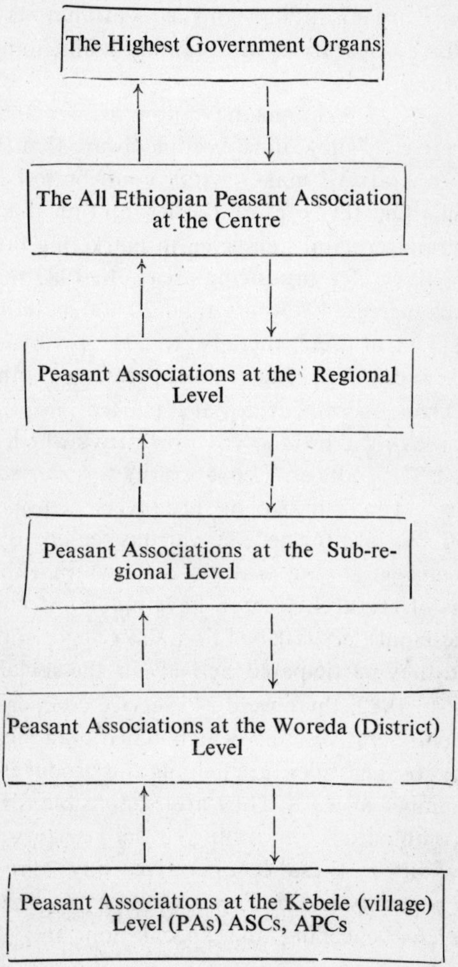

Source : Mengistu Woube, Problems of Land Reform Implementation in Rural Ethiopia, A Case Study of Dejen and Wolmera Districts, Geografiska Regionstudier Nr. 16, Uppsala, 1986, p. 73.

ties will not last long.[9] Farmers are generally convinced of the advantages of the co-operatives by actions and examples. In an effort to encourage these co-operatives, the Government is providing necessary assistance to the needy co-operatives. It gives not only organisational and educational assistance but also helps the co-operatives in obtaining loans, improved seeds and agricultural tools.

The people of Ethiopia have now realised the advantages of co-operatives. They have become aware that the establishment of co-operatives makes vital contribution for socialist construction. On the eve of the Revolution there were only 150 co-operatives mainly engaged in marketing predominantly located in the coffee producing areas. In 1982 the number of co-operatives increased to more than 20,000 including all types of societies. Out of these, there, were 593 service co-operatives (recognised) and 837 producers' co-operatives (both Malba and Wolba).[10] The Government has also started granting of legal status to some of the producers co-operative which fulfil certain conditions. 57 producers' co-operatives have acquired this legal status. The number of producers' co-operative (both Malba and Wolba) formed in various regions is encouraging and the latest region-wise position is shown in Table VIII.1.

Peasants of Hararghe region have made immense sacrifices. On the one hand they crushed Somalian aggressors and on the other hand they participated actively in the socialist construction efforts. In 1982, there were 245 service co-operatives and 94 producers' co-operatives, of which 6 acquired legal position.[11] The service co-operatives are helping the producers' co-operatives in a number of ways. They are responsible for purchasing industrial commodities and selling to the peasants at reasonable prices. In addition, these co-operatives buy agricultural products from peasants and in turn sell them to the town consumers thereby checking the speculation and exploitation.

9. Information and Propaganda Committee of the Fifth Anniversary of the Ethiopian Revolution, Ethiopian Revolution Fifth Anniversary, Addis Ababa, 1979, p. 26.
10. The Ethiopian Herald, Addis Ababa, 9 April, 1982.
11. The Ethiopian Herald, Addis Ababa, 26 March, 1982.

TABLE VIII.1

Number of Producers' Co-operative with Membership and Area

Province	Producers' Co-operatives	Membership	Area (ha)	Legal Status
Hararghe	94	4995	7116	6
Shoa	86	5182	14653	7
Arussi	85	3553	8137	6
Gojam	83	2765	13673	10
Bale	82	4695	8689	4
Wollega	76	5188	24686	N.A.
Wollo	71	5194	5840	2
Illubabor	71	3781	4560	3
Sidama	69	12008	14105	2
Kafa	50	2269	11445	3
Gondar	43	1963	3358	6
Gamu-Gofa	16	1881	3141	6
Tigre	8	230	351	2
Eritrea	3	88	20	—

Source : The Ethiopian Herald, Addis Ababa, 9 April, 1982.

These co-operatives also purchase implements (Spade, Pick axes, etc.) from distributing agencies and sell to producers' co-operatives. In future, they have plans to purchase modern agricultural equipments like tractors and supply them to the producers' co-operatives. Service co-operatives also give loans to the producers' co-operative without interest.

The available information shows that 61 per cent of the producers' co-operatives are engaged in the grain production and the rest in the production of other crops. Recently, in order to encourage the establishment of more producers' co-operative, the authorities have started giving prizes if the co-operatives fulfil two conditions. Firstly, the extent to which the co-operatives abide by the principle "from each according to his ability to each according to his work" and secondly, the degree to which co-operatives carry out their duties according to plans and objectives.

The transformation of agriculture also requires the establishment of large scale state agricultural farms. The necessity of having state farms side by side with co-operatives is due to the fact that socialist Ethiopia inherited privately owned large scale farms and plantations from the old regime. These privately owned farms were controlled by big capitalists. Such farms have now come under the state ownership and form the nuclei for large scale state farming. The objective of state farms is not merely to produce raw material and crops required for domestic industrial consumption and foreign trade but also to meet the foodstuff needs of the urban centres. Another important objective is to speed up the transformation of agriculture which occupies such a prominent place in the economic life of the nation by introducing better farming technology, by supplying choice seeds and, in general by setting example. Accordingly, a very high priority has been given to the expansion and strengthening of state farms. The size of the state farms has increased from 73,000 hectares in 1979 to 300,000 hectares in 1982.[12] These farms are chiefly being developed in the Awash Valley, the Awash-Shashamanne area, Arba-Minch, Chilalo, Bale Valley and Eritrea. At present, the major attention is being paid to establishing state farms in Gemu-Gofa, Bale and Kefa provinces.

Agricultural Growth

The 1975 Proclamation of Agrarian Reform has no doubt resulted in a radical structural change by creating three important sectors—Peasantry, Co-operatives and State Farms. The main objective of the reform is the emancipation of the peasant from poverty and the general improvement of the standard of living of the whole working people. This can be possible only by increasing agricultural production. Despite the fact that the agrarian reform was a vital step toward social and economic advancement in the long run, it caused a dramatic drop in agricultural production in recent years.

---

12. Documents and Resolutions of 2nd COPWE Congress, Addis Ababa, January, 1983, p. 41.

TABLE VIII.2

Estimated Production and Area Under Major Crops

Crop	Total Production (000 qt)				Percentage Change			
	1975-76	76-77	77-78	78-79	1976-77 over 1975-76	1977-78 over 1976-77	1978-79 over 1977-78	1978-79 over 1975-76
Cereals	46986	43398	39746	41521	−7.6	−8.4	−4.5	−11.6
Pulses	4619	6241	5148	4707	35.1	−17.5	−8.6	1.9
Oilseeds	1300	725	867	844	−44.2	19.6	−2.7	−34.7
Industrial & Other crops	533	687	727	792	28.9	5.8	8.9	48.6
Total	53439	51052	46490	47865	−4.5	−9.0	3.0	−10.4
	Area Under Crope (000 ha)							
Cereals	4553	4376	4617	4857	−3.9	5.5	5.2	6.6
Pulses	646	669	629	652	3.6	−6.0	3.6	0.9
Oilseeds	367	231	214	237	−37.1	−7.5	11.0	−35.4
Industrial & Other crops	26	29	30	34	18.0	1.7	15.3	36.0
Total	5592	5307	5491	5780	−5.1	3.5	5.3	3.4

Source : National Bank of Ethiopia—Annual Report, 1979, Addis Ababa.

The radical nature of the land reform programme struck the economy negatively once the programme began to be implemented seriously. Table VIII.2 shows that the total agricultural output which was 53.4 million quintals in 1975-76 decreased to 51.0 million quintals in 1976-77 or 4.5 per cent decrease just in one year. The production further decreased by 9.0 per cent in 1977-78. For the first time since the Revolution the total agricultural output increased by 3 per cent in 1978-79. During the entire three year period, the fall in the agricultural output was about 10.4 per cent or 3.4 per cent per annum. Consequently, Ethiopia had a grain deficit of 200,000 tons in 1976-77 which increased to 468,000 tons in 1977-78 but fell to 270,000 tons in 1978-79. On food shortage, Lt. Col. Mengistu himself called the situation "frightening" during a speech in September, 1978.[13]

The overall agricultural production for the whole country is classified into cereals, pulses, oilseeds and industrial and cash crops. In 1975-76 the production of cereal crops was 46.9 million quintals which decreased to 43.4 million quintals or 7.6 per cent in 1976-77 and further by 8.4 per cent in 1977-78. The performance of cereal output in 1978-79 showed some improvement and the economy registered 4.5 per cent increase over the previous year. The break-up of the cereal crops shows that the production of wheat, barley and teff rose by 7.6 per cent, 2.0 per cent and 6.6 per cent respectively in 1978-79. Due to high domestic prices, much emphasis was placed upon the production of cereals by the farmers. Consequently, the production of pulses and oilseeds decreased. In 1976-77 the production of pulses of course increased by 35.1 per cent but in the subsequent two years the output fell by 17.5 per cent and 8.6 per cent respectively. The production of oilseeds decreased by about 12 per cent per annum during the period under review. The only systematic increase in the production was in the case of industrial and cash crops. The annual growth rate of these crops was 16 per cent. Coffee had been doing well in terms of value if not volume.

13. African Research Bulletin, Economic, Financial and Technical Series, August 15-September 14, 1978, p. 4812.

Table VIII.2 also shows the area under important crops. There was some positive correlation between the area under cultivation and production of agricultural crops between 1975-76 and 1976-77. The total area under crops decreased by 5.1 per cent while the total production of crops decreased by 4.5 per cent. Another important period where the positive correlation is observed was between 1977-78 and 1978-79. During this period the area increased by 5.3 per cent and the output went up by 3 per cent. Negative correlation was observed between 1976-77 and 1978-79. Area increased by 3.5 per cent while the output decreased by 9.0 per cent. But it is clear that in all the cases the contribution of yield was negative. Same trends have been observed in the case of cereals. During the entire three year period, the area under cereal crops increased by 6.6 per cent while the total output decreased by 11.6 per cent. The area and production of oilseeds also moved in the same direction. The area decreased by 35.4 per cent and the production fell by 34.7 per cent. The contribution of yield was positive in the case of pulses and industrial and cash crops. The area under pulses increased by one per cent but the production went up by two per cent. As such the area and yield contributed more or less equally. In the case of industrial and cash crops the area increased by 36.0 per cent and production went up by 48.6 per cent.

Peasant and Co-operative Sectors

Agricultural activity is divided into peasant sector, co-operative sector and state farm sector. The crop production of peasant sector constitutes about 94 per cent of the country's overall crop production and about 97 per cent of the total cultivated area.[14] Table VIII.3 reveals that the production of agricultural commodities of peasant sector which was estimated at 52.0 million quintals in 1975-76 decreased to 45.1 million quintals in 1978-79. Correspondingly, the area under production increased from 5.4 million hectares to 5.6 million hectares during the same period. This is an increase 3.7 per cent in

14. National Bank of Ethiopia, Annual Report, Addis Ababa, 1979, p. 11.

TABLE VIII.3

Peasant Holdings-Estimates of Area and Production of Major Crops

Crop	1975-76		1978-79		1978-79 Over 1975-76	
	Area	Production	Area	Production	Area	Production
Cereals	4488	46186	4738	39663	5.5	—14.1
Pulses	642	4566	646	4679	0.6	2.5
Others	352	1267	209	766	—40.0	—39.5
TOTAL	5485	52018	5594	45108	3.7	—13.3
Co-operatives-Estimates of Area and Production of Major Crops						
Cereals	46.0	381	36.0	290	—21.7	—23.8
Pulses	5.5	41	2.8	18	—49.1	—56.1
Others	10.0	22	7.2	22	—28.7	—
TOTAL	61.6	444	46.0	330	—24.5	—25.7
State Farm-Estimates of Area and Production of Major Crops						
Cereals	18.5	419.5	89.6	1567.9	384.3	273.7
Pulses	1.8	12.2	2.4	10.4	33.3	—14.7
Oilseeds	0.2	0.5	19.5	46.4	9650.0	9180.0
Industrial and Cash Crops	25.0	533.6	34.6	792.6	38.4	48.5
TOTAL	45.5	965.8	146.1	2417.3	221.1	150.3

Production in '000 quintals and Area in '000 ha.

Source : National Bank of Ethiopia, Annual Report, 1979, pp. 12 and 14.

area as against the fall of 13.3 per cent in production during the period under review. The production of cereals which represents about 88 per cent of the total production of peasant sector went down by 14.1 per cent while the area under crops rose by 5.5 per cent. It was only in the case of pulses where the production rose by 2.5 per cent and area increased by 0.6 per cent during the period under review.

As far as the co-operative sector is concerned, the levels of output and area under crops went down more or less in the same percentage. The crop output from this sector decreased by 25.7 per cent while the area under crop fell by 24.5 per cent during the period under review. The production of cereals which constitutes about 88 per cent of the total produced by the co-operatives is still less than one percent of the overall country's cereal production. However, the future policy of the government is to establish co-operative farms in the country by converting the existing peasant associations into producers' co-operatives.[15]

State Agricultural Farms

Agricultural production from the state sector constitutes about 5 per cent of the country's overall crop production. Its farms account for about 1.5 per cent of the nation's total land under cultivation. Although crop production from the state farms may also be used to supplement local grain consumption, its main objective is to provide agricultural products for export and local industrial inputs.[16] The crop production from this sector has shown a remarkable increase. In 1975-76 the total crop production of this sector was about 965.8 thousand quintals which increased 2417.3 thousand quintals in 1978-79. In terms of percentage it comes to about 150 per cent. The total farm land cultivated under this sector has also shown a remarkable growth rising from 45.5 thousand hectares to 146.1 thousand hectares or by 221 per cent during the period under review. The cereals and industrial and cash crops constitute the bulk of the produce of this sector. The annual

15. *Ibid.*, p. 13.
16. *Ibid.*, p. 11.

rate of growth of these two types of crops was 91.2 per cent and 16.1 per cent respectively during the same period. This shows that the state farms also play an important role in the production of cereals to supplement the domestic food requirements of the country. Although oilseeds and pulses are the most important export crops next to coffee, they do not constitute more than 2.3 per cent of the overall crop production of the state sector. The production of oilseeds has also shown a satisfactory progress. The annual production of this crop was virtually negligible in 1975-76 and it rose to 46 thousand tons in 1978-79. The production of pulses is also picking up since 1976-77.

During the pre-revolutionary period, policies of the government in general favoured expansion of large scale commercial farming and failed to help the peasant agriculture. In fact the promotion and rapid expansion of large commercial farms made peasants more vulnerable to famine. Inspite of the warning of Assefa and Chole that the underdeveloped state of agriculture and the government indifference to improving agricultural productivity of small farmers would lead to widespread food shortages in the near future,[17] the government did not show any interest in the progress of peasant sector. It never worked seriously to improve the productivity of land of small farmers.

The present government has of course changed the policy by shifting priority to the development of peasant farming for increasing the production of food crops. It launched Agricultural Minimum Package Programme in 1979 at a cost of US $ 77 million in order to improve the productivity of land of peasant farmers over two years by providing fertilisers, improved seeds and other agricultural implements.[18] The programme also proposed to ensure long term success through soil and water conservation. The statistical data show that the programme has significant impact on the production of agricultural crops. For example, in 1980 the total production of agri-

17. Willene Johnson "Food and Politics : A Case Study of Ethiopia" Horn of Africa, Vol. II, No. 2, January-March 1979, p. 31.
18. The Economist Intelligence Unit, Quarterly Economic Review of Ethiopia, Uganda and Somalia, 1st Quarter, 1981, p. 13.

culture was estimated at 56.0 million quintals as against 47.6 million quintals in the previous year thus accounting for 16.9 per cent increase.[19] This production figure has surpassed all the previous records. But if the entire period since the Revolution is taken, the annual rate of growth of agricultural crops does not exceed 1.2 per cent.

In 1980, Ten-Year Investment Programme was launched. To this end, the Programme sets out certain goals to be achieved by 1990 *i.e.* to double the gross domestic product in real terms and to increase agricultural production by 60 per cent alongwith targets for other sectors. Theorertically and technically the Ten-Year Investment Programme is a sound policy statement. Whether it will be able to achieve these goals is a question that for the moment cannot be answered.[20] As far agricultural sector is concerned, the performance of this sector has been far from satisfactory. Similarly, according to the available information based on the output, area and yield figures compiled by the CSO, Table VIII.4 shows that during the period from 1979-80 to 1981-82, the output of agriculture decreased by 16.4 per cant. The maximum decrease was observed in the case of oilseeds (34.9 per cent) followed by cereals by 18.8 per cent. This decrease in agricultural production was mostly due to decrease both in area under cultivation and yield per hectare. The production of cereals and pulses in 1983 and 1984 was no more than in 1972—the

TABLE VIII 4

Estimated Production of Major Crops (000 Qt.)

Crop	1979-80	1980-81	1981-82	1981-82 over 1979-80
Cereals	65676	36119	53285	—18.8
Pulses	9336	8167	8010	—14.2
Oilseeds	1655	1330	1077	—34.9
Total	74668	65617	62373	—16.14

Source : CSO, Area, Production and Yield of Major Crops, August, 1982

19. National Bank of Ethiopia, Annual Report, 1980, *op. cit.*, p. 8.
20. Peter Schwab, Ethiopia-Politics, Economics and Society, *op. cit.*, p. 82.

year of the Great Famine.[21] It is clear that the situation is potentially devastating with production remaining at almost historic lows in recent years.

Conclusion

According to the Government documents the slackening or inability to strengthen discipline among peasants is an important cause for the decline in agricultural production. The organisation of productive forces in the agricultural sector and the type of agricultural implements which they employ are the other causes limiting the contribution to increase production. "It will be difficult to raise the productivity of the agricultural sector unless the average small and scattered holdings of the peasantry, the limited use of fertilisers, choice seeds and insecticides as well as the backward farm implements are changed and improved."[22]

Some of the economists even feel that the creation of small farms without much improvement in farming practices will tend to freeze Ethiopian agriculture at the subsistence level. Though the Peasants' Associations were considered to be instrumental in bringing fundamental changes in agricultural practices, their effectiveness remained highly uneven. The Government grants broad functions to these Associations but it withholds the economic resources needed to make their power effective.

A struggle has been of course waged over the past years for the expansion of producers' co-operatives in the countryside. A good number of producers' co-operatives at different stages of development have been set up. But the achievements of the past three years are inadequate in view of the immense tasks before the Government. The number of such co-operatives is not very large. Though the directive issued elaborates in detail the procedures and methods of planning and management of funds, the implementation requires training of members. Thus the effort so far expanded by way of credit grants, material support and related encouragement to them does not measure up to the priority.

21. *Ibid.*, p. 80.
22. Documents and Resolutions of the 2nd COPWE Congress, *op. cit.*, p. 39.

The performance of state farms has also not been satisfactory. For various reasons, productivity of state farms has not yet reached a satisfactory level. The yield of state farms should be an average of 25 quintals of wheat per hectares. The results so far achieved do not exceed a yield of 14 quintals per hectare. Even the yield of main crop (maize) produced by the state farms is hardly higher than that of peasant holdings. These farms were developed hastly without much reliable studies. It was even found necessary in the course of fourth and fifth programmes to check the expansion of such farms while at the same time attempting to consolidate these farms in manpower, equipment, finance and other facilities.[23]

The above analysis clearly shows that nothing but the very initial phase of land reform has been overcome so far. The reform did eliminate one of the major structural hindrances of agricultural development. A large part of the rural population can now retain for themselves the fruits of their labour. But that is not enough. The mere redistribution of land and the settlement of people in new areas will not by themselves achieve the land reform objectives. The present land reform must be regarded as an integral part of the overall developmet plan. The success of the programme depends upon the willingness, firm determination and commitment of the Ethiopian Government and the peasants.

23. *Ibid.*, p. 41,

CHAPTER NINE

Implementation of Land Reform

SINCE the Ethiopian Revolution, six major land reform proclamations have been promulgated with three main objectives : the nationalisation of all rural land and the distribution of land to the tillers upto 10 hectáres, the decentralisation of power and the formation of co-operative societies in order to lay the basis for socialism. These proclamations have been regarded as quite radical because they have brought about land distribution and organisational changes in Ethiopia.[1] Such changes give the landless people rights over land.

In order to carry out the reform programme and to strengthen the political consciousness of farmers, the Government launched its "Development Through Co-operation Campaign," known as the Zemecha Programme. The Government selected about 56,000 Zemecha students who were sent out to the rural areas for the politisation of farmers, the eradication of illiteracy and the teaching of health care in the rural areas.[2] An amount

1. Mengistu Woube, Problems of Land Reform Implementation in Rural Ethiopia, A Case Study of Dejen and Wolmera Districts, *op. cit.*, p. 51.
2. *Ibid.*, p. 52.

of 50 million Birr was sanctioned for this purpose. But initially enormous problems did arise out of the implementation of the land reform. First of all, the Zemecha Programme had contradictory effects in the rural areas. The students sent out soon clashed with the police and administrators. The latter might have had land for themselves or might have looked forward to receiving some land in future. In such clashes, according to Hailu W. Michael, about 92 students died in 1976.[3] The Zemecha Programme was very costly to the country both in terms of human lives and material resources. In the early 1976, most of the students had been recalled to the cities. The Proclamation of December 1975 was designed to establish a more permanent mobilising structure in the rural areas and the Zemecha students, whose usefulness to the PMAC was now exhausted, brought back to the cities a deep suspicion of policies of the PMAC.[4] It fuelled the conflict that had already broken out there between the ruling party and its opponents.

Although no reliable research work on the implementation of the Proclamation on land reform is available, it can be said that substantial progress had been made in those areas of the country where there were large scale holdings, extensive patterns of tenancy and absentee landlordism. But the implementation of land reform Proclamation was more difficult in share cropping areas. In South, the Oromo peasantry seized the lands and in some cases even killed the former landlords. Since the Oromo peasantry had not perceived themselves before the Revolution as owning the land, they were not threatened by the new system.[5] In large parts of Begemdar, Tigre and Wollo and to a lesser extent in other central provinces the land reform Proclamation did not correspond to the needs of the peasantry. The traditional forms of land tenures were mostly communal and precluded the confiscation of large estates and as such the pro-

3. Hailu W. Michael, Zemecha, An Attempt to Rural Transformation in Ethiopia (Draft), Addis Ababa April 15, 1975, p. 4.
4. Fred Halliday and Maxine Molyneux, The Ethiopian Revolution, British Library, London, 1981, p. 106.
5. *Ibid.*, p. 105.

clamation was not relevant. Similarly, in those areas where the system of 'rist' land had prevailed and provided a form of ownership to the labourers, the results were less favourable. In the Afar areas, the Sultan was able to mobilise resistance among his people, since in this Muslim area of the north-east unlike Oromo, both land owners and peasants shared a single religious affiliation. Resistance broke out in many parts of the north and it effectively substracted much of the areas from the Government control during the initial period of the land reform programme.[6]

The PMAC was of course not the first to call for land reform in Ethiopia, but certainly it was the Military Government which implemented a land reform in a much more radical manner. It went against the large number of landowners and even against the wishes of the very layer of officers from whom the Dergue had emerged. They lost their traditional right to grants of land in return for military services. Most of them sent an ultimatum to the Dergue which included a demand that soldier's land be exempted from the Proclamation of land reform. These developments also created problems for the Government in implementing the land reform.

The most important aspect of the Proclamation is the division of implementation responsibilities into vaguely defined compartments between peasant associations and the bureaucracy.[7] It completely by-passed local administrators under the Ministry of Interior. This was of course done to dismantle feudal administrators, but there was considerable tension among them. It is true that the government is ideologically committed to respecting peasants and rely on them for the implementation of land reform and other various programmes locally. The government wants to trust them, but fears that it cannot do so. It grants broad functions to peasant associations, but does not provide economic resources needed to make their power effective.[8]

---

6. *Ibid.*, p. 105.
7. Paul Beiretzke, Land Reform in Revolutionary Ethiopia, *op. cit.*, p. 653.
8. *Ihid.*, p. 654.

Structural Problems

The most important step in the land reform process was the establishment of peasant associations. These associations were given the functions of distribution of land, solving disputes between members, carrying out co-operative undertakings and electing committees to supervise these tasks. In a subsequent Proclamation, the peasant associations were given legal status, their main responsibilities being to maintain law and order within their respective areas. But these associations faced a number of problems at the time of distribution of land to the peasants. In some areas due to less land available, it was difficult to redistribute sufficient land to all peasants. In a case study of Dejen district, it has been observed that most of the peasants possessed less than 5 hectares in pre-reform period. The 10 hectares ceiling announced by the Proclamation of 1975 affected only a small number of peasants. Since arable land was small compared to population density, the peasants could not get 10 hectares of land as stated in the Proclamation.[9] Some cultivable land was available for distribution but such land was required for grazing. The peasant associations in Dejen district redistributed only the available land which had existed before the land reform. Consequently, the size of landholdings has declined substantially and in recent years the number of landholders who possess less than 4 hectares of land has increased.

Similarly, in densely populated areas where all cultivable land was taken into production, redistribution of land also created a difficult political problem. For example, one peasant association in Begi Woreda had decided to redistribute all land so that each member could get equally big holdings. The political decision was approved by a broad majority within the association. Nevertheless, the results proved so discouraging that the association abandoned the whole idea. This was due to absolute land scarcity prevailing within the association's area. There had been no big landlord whose land could be confiscated and portioned out to the members. In this case, redistribution means that some plots of land were taken from

9. Mengistu Woube, Problems of Land Reform Implementation in Rural Ethiopia, *op. cit.*, p. 85.

the less poor and given to the poorest. The peasant association leaders concluded: "Instead of having one poor and one middle farmer, the redistribution resulted in two poor farmers."[10] The association in Heena resolved that the principle for re-distribution of land was not to be absolute equality. Attention was instead to be paid to family size so that big families should get more land than small families.[11]

The principle of equal re-distribution of land is a desirable goal but difficult to translate into practice due to various reasons. Since the shape, quality and number of parcels of land varies within the region, it becomes difficult to re-distribute land to each member equally. Moreover, the relationship between household size and land area is also changing. The number of dependents in many households increases because of marriages or births before the next redistribution of land takes place.[12] This is one of the important reasons why in some households there are more dependents and less land than in others.

Further, boundaries of the peasant associations were delineated without much knowledge about the topographic nature of the land and the number of people living in that area. In demarcating the boundaries, the committees of the associations followed lines such as roads, rivers, mountains, etc. This type of approach created conflicts and inequalities in the size of holdings. In 1983, a large number of peasant associations in Dejen district demanded the re-delineation of boundaries and the redistribution of land on an equal basis. But such demands could not be accepted due to lack of various studies and adequate financial and material resources.[13]

Peasant associations are expected to co-ordinate administrative functions, agitate and mobilise the people to participate

10. Michael Stahl, New Seeds in Old Soil, A Study of Land Reform Process in Western Wollega, Ethiopia 1975-76, Scandinavian Institute of African Studies, Research Report No. 40, Uppsala, 1977 p. 33.
11. *Ibid.*, p. 55.
12. Mengistu Woube, Problems of Land Reform Implementation in Rural Ethiopia, *op. cit.*, p. 88.
13. *Ibid.*, p. 79.

in political and economic activities and maintain the security of the region.[14] In some areas these associations are really participating in building and running elementary schools. They also take part in the ongoing literacy programme. Some of the associations run well-established consumer shops, provide funds for the newly established agricultural producers' co-operatives. But such associations have not been able to generate new ideas as to how the rural sector could be developed.[15] It is a fact that peasant associations have emerged as the new local institutions in the countryside and they have replaced the feudal institutions as links between individual farmers and the administration. But due to their 'appendage' status their participation in decision-making and implementation of administrative decisions is nominal.[16]

Any member of the peasant association who is 21 years of age and does not own more than 10 hectares of land has the right to be elected as a member of the executive committee. Once elected, the members of the associations have no power to dismiss inefficient leaders. They cannot even criticize corrupt administrators. Such elected members can be removed only by the approval of the government. As a result of this, members do not take much interest in exposing corrupt people in the open because their activities could be regarded as anti-revolutionary elements.[17] They are not even free to express their views in meetings.

The general assembly of all the members is the highest decision making body of the peasant association. The chairman of the general assembly is the chairman of the executive committee. It is the duty of the chairman and the secretary of the committee to walk from village to village in order to settle disputes of the members. But unfortunately they receive no salary or

14. Alula Abate, Peasant Associations and Collective Agriculture in Ethiopia, Erdkunde, 1983, p. 118.
15. Mengistu Woube, Problems of Land Reform Implementation in Rural Ethiopia *op. cit.*, p. 79.
16. Alula Abate, Peasant Associations and Collective Agriculture in Ethiopia, *op. cit.*, p. 188.
17. Mengistu Woube, Problems of Land Reform Implementation in Rural Ethiopia, *op. cit.*, 80.

wages. Like others, they are young or middle-aged farmers who can support their families with their own production. Because of their duties in the executive committee of the peasant association, they have little time to cultivate their land for their own livelihood and their families generally suffer from a shortage of agricultural output. "Executive committee members of the KPAs do not have sufficient time for their own family activities, because they are engaged in the execution of directives and orders of government almost with no time limit. They cannot satisfy the needs of all members in the association or the desires of the bureaucracy."[18]

Service Co-operatives

One of the functions of peasant associations was to establish marketing and credit co-operatives and other associations like the *debo* which would help farmers to co-operate in manual and other works.[19] Accordingly, agricultural service co-operatives have been established by a large number of peasant associations. But a study of Dejen district shows that these service co-operatives are not functioning properly. Basic commodities and agricultural implements are not available in most of the service co-operatives. The co-operative shops function poorly and peasants are not protected from profiteering merchants. The main cause of this malfunctioning is that the agricultural service co-operatives are heavily dependent on government institutions due to shortages of trained personnel, funds and proper transportation facilities.[20] The Ethiopian Domestic Distribution Corporation is supposed to provide basic commodities to the co-operative shops on priority basis. But due to an inefficient distribution system, the EDDC is not in a position to supply commodities in time. Moreover, peasants are scattered and generally live far from the co-operative shops and as such they cannot avail of the facilities of the shops. These factors

18. *Ibid.*, p. 81.
19. Proclamation No. 31 of 1975, Chapter 3,
20. Mengistu Woube, Problems of Land Reform Implementation in Rural Ethiopia, *op cit.*, p. 103.

contribute to the unsatisfactory services of many agricultural service co-operatives.[21]

The traditional system of credit was abolished by the Proclamation of 1975. The peasants can get fertilisers on credit by making a 15 per cent down payment and the rest must be paid just after the harvest. But the problem is that the members cannot purchase fertilisers directly from their co-operatives. It has to be provided through different channels. For example, a list showing the amount of fertilisers required by members for the coming season is collected by the chairmen of the agricultural service co-operatives/peasant associations and is handed over to the chairman of the Woreda Peasant Association who in turn gives it to the Ministry of Agriculture (Woreda Office), then to the Agricultural Marketing Corporation and finally to the Agricultural Industrial Development Bank. Due to this long delay in the supply process, less fertiliser is used on individual farms. For example, in Dejen district about 68 per cent of all farmers in the pre-reform period used fertilisers whereas in the post-reform period only 57 per cent did so.[22]

The farmers can purchase fertilisers on credit basis on the condition that as soon as the crop is harvested they will sell it to the Agricultural Marketing Corporation in order to pay back their debt. If repayment of loan is not done in time, the farmers will not be able to receive fertilisers on credit for the coming season. But unfortunately, the prices set by the Corporation for different types of grains are very low. During the period from 1980 to 1983, the Dejen district study shows that the price of teff increased by less than 20 per cent whereas the price of fertiliser increased much more than this. As a result of this, the farmers could not pay their debts from the previous year.[23]

In the grain marketing field the specific aims of the government are to assure stable producer and consumer prices, maintain adequate producer incentives, reduce marketing margins

21. *Ibid.*, p. 103.
22. *Ibid.*, p. 104.
23. *Ibid.*, p. 104.

through greater efficiency and assure an adequate food supply in all parts of the country.[24] In order to achieve these objectives, farmers are obliged to sell at a fixed price a fixed quantity of grain to the Corporation. The quantity and price of grain is decided by a higher government body. Peasants sell their respective quota within a limited period to the co-operatives. Finally, the Agricultural Marketing Corporation collects the grain from the co-operatives and pays 5 Birr per quintal to the co-operatives for storage and personnel services rendered by them. This system has helped the farmers in many ways. Under this system, farmers know in advance when and where they are going to sell and how much price they are going to get. It also saves the farmers from the middlemen. But the quota system has also created a number of problems for the peasants. The peasants who do not have enough grain to sell are the real sufferers. In the case of Dejen district, it has been observed that 30 per cent did not have enough grain to sell in 1983. They bought some grain even by selling their domestic animals and sold the required quintals of grain to the co-operatives. It means that all peasants have to sell either grain or animals in order to purchase basic commodities, agricultural inputs and to pay taxes. The study further reveals that only 36 per cent farmers wished to sell their grain to the Corporation because of the low price paid by it. On the other hand, about 52 per cent wanted to sell to individual consumers and remaining 12 per cent to middlemen. This means that the Corporation which tries to derive profiteering merchants out of the grain market has not been able to do so. The prevailing system will continue as long as farmers are not provided better prices for their products.[25]

Producers' Co-operatives

The Proclamation has of course placed land in the hands of peasants. But the size of average holdings is very small. Peasants have very little farm equipments and draught animals. The

---

24. John Holmberg, Grain Marketing and Land Reform in Ethiopia, *op. cit.*, p. 12.
25. Mengistu Woube, Problems of Land Reform Implementation in Rural Ethiopia, *op. cit.*, p. 106.

main objectives for the establishment of agricultural producers' co-operatives are to free the small holding farmers from the poverty and to make them supporters of socialism.[26] When the agricultural producers' co-operative is registered officially, it takes over the role of first the peasant association and then the service co-operative. This is a simple form called Melba co-operative which permits only partial joint use of farming tools and draught animals. It gradually moves to the second stage of collectivisation called Wolba co-operatives and finally to the most advanced stage called Woland.

It is the policy of the government to give top priority to the establishment of producers' co-operatives than to the individual farms. The co-operatives are provided a greater quantity of fertilisers and improved seeds at a reasonable price. It is true that the producers' co-operatives still use traditional methods for cultivation. But the division of labour between men and women according to age and interest helps the co-operatives to organise labour in a disciplined manner.[27] Heavy work like ploughing, winnowing, etc. is done by strong men and medium work like replanting seedlings, etc. is mostly done by women. In theory, all members of the co-operatives are supposed to contribute labour.

Peasants on the private farms, even when grouped in peasant associations, have the freedom to choose what to grow, whether or not to sell all their produce. But peasants on the agricultural producers' co-operative farms do not enjoy the same right. They have to follow strict directives issued by the Ministry of Agriculture for the production of a given quantity of grain for sale. A recent study[28] conducted in the case of Wolmera district shows that "the majority of both individual and co-operative farmers in Wolmera are not much attracted by the APCs. The APCs do not have personnel and the capital to create attractive and catchy activities. Peasants join co-operatives mainly because they lack sufficient means of production and face the problems of shortage of a relatively constant income. Advanced or

---

26. The Ethiopian Herald, January 29, 1982.
27. Mengistu Woube, Problems of Land Reform Implementation in Rural Ethiopia, *op. cit.*, p. 113.
28. *Ibid.*, p. 152.

model co-operatives... do not exist in Wolmera. Therefore peasants do not perceive any real incentive for joining the APCs. Even among co-operative members, most of the members care more for their own garden crops than for the collective farm operation." The study further reveals that the membership of the existing producers' co-operatives has declined in recent years because the co-operatives have no attractive agricultural methods that give tangible results in a short period, those who work hard and those who work less hard receive equal shares of the production and many farmers prefer to stay under the service co-operatives rather than to advance to producers' co-operatives and go through collectivisation.

But the producers' co-operatives have proved that members of different religions coming from different ethnic groups can live together. Discrimination between high and low-caste peasants is reduced. Members are relatively better fed than the individual farmers. So the producers' co-operatives have greater potential for further development than the individual farms. But most of the individual farmers are not aware of the advantages of the co-operatives. Whether or not more individual farmers will join the co-operatives will essentially depend on the ability of the leaders of co-operatives and the support given by the government.[29]

Distribution of Farm Oxen

Before the Revolution, cattle were accumulated by landlords, rich farmers and even by civil servants. The concentration of cattle-ownership to the ruling classes meant that many peasants lacked oxen and had to rent them. Rent of draught-oxen constituted an important element in the exploitation of peasants by landlords. Agricultural surveys conducted in the 1960's indicated that about two-thirds of the peasantry were not self-sufficient in means of production and they were dependent on other people for ploughing. A large number of peasants owned only one ox. A survey of five Woredas in Ghimbi reported that one-third of peasants did not own any draught-oxen while another one-third of the peasants owned only one ox.[30]

29. *Ibid.*, p. 117.
30. Michael Stahl, New Seeds in Old Soil, *op. cit.*, p. 39.

The 1975 Proclamation brought important changes to this pattern. It states that the tenant "shall have the right to retain agricultural implements and a pair of farm oxen belonging to the land owner for which a reasonable compensation shall be paid to the landowner within a period of not exceeding three years ; provided that a landowner with no other agricultural implements and farm oxen or no adequate means of livelihood may take back his implements and oxen from the tenant."[31] When the peasants took over the land and property from landlords cattle were also taken. The confiscated cattle were put under the ownership of the peasant association and thereafter given on credit to members who lacked draught-oxen.

But in many areas landlords managed to hide their cattle before the peasants took action. When land reform was proclaimed, large herds of cattle were reported to have crossed the Sudan border. Others were hidden in the Didessa valley. Slaugther houses in the small towns also received more cattle than usual. Due to all these reasons, the number of draught-oxen for the peasants decreased.[32] When the peasants got control over the land, they lacked oxen to plough it. The government at the time of announcing land reform had not foreseen the ox-problem. However, in the subsequent years, the number of draught-oxen had increased and peasants were provided oxen for cultivation purposes. But there is still shortage of oxen.

The lack of oxen has created conflicts within the associations, between those peasants who have access to oxen and those who have not. The Proclamation of 1975 only sanctioned confiscation of cattle belonging to landlords. Oxen belonging to peasants were to remain the private property of the previous owners. Inequalities therefore remained inside the peasants associations.[33] In other words access to oxen continues to be a burning issue for peasant associations in almost all the districts.

The draught animals kept within a peasant association

---

31. Proclamation No. 31 of 1975, Chapter 2.
32. Michael Stahl, New Seeds in Oil Soils, *op. cit.*, p. 39.
33. *Ibid.*, p. 42.

continue to be private property. But their sale can proceed only within the association between members. If a member wants to sell cattle outside the association, he has to apply to the executive committee for permission. But the committee generally turns down such applications. It may be mentioned here that those who have oxen for sale are rich peasants or middle peasants. Since the committee refuses sale of oxen outside the association and the members are poor and cannot buy an ox, those who have surplus oxen retain them. Consequently, such surplus-oxen usually remain unproductive in agricultural work.[34]

In the case of Ethiopia land reform cannot make it possible for farmers to replace farm oxen by tractors within a short period of time. "Not only does this require huge amounts of capital and highly qualified personnel, but it is alien to the introduction of intermediate agricultural technology. The demand for farm oxen is much higher than the supply and this is not only a local but a national problem. In order to solve this acute problem, relatively simple and inexpensive measures could be taken, such as, systematic introduction of better breeding and the provision of sufficient fodder and proper care and the introduction of an appropriate intermediate agricultural technology."[35]

Land Fragmentation

During the pre-revolutionary period, the land use-rights brought extreme fragmentation of land. Peasants were able to claim land through several sources of descendency and they could claim land in any of the different areas. This also resulted in social tension and conflicts among peasants which led to many court cases. The land fragmentation problems became even more pronounced after the reform. A survey conducted in Dejen shows that in the pre-reform period 61 farmers out of 200 owned 3-4 parcels, but after the reform, the corresponding number was 125. Similarly, the number of holders of 5 or more parcels was 30 but after the land reform

34. *Ibid.*, p. 42.
35. Mengistu Woube, Problems of Land Reform Implementation in Rural Ethiopia, *op. cit.*, p. 121.

this figure had decreased to 12. Land fragmentation has been affected by land redistribution and the resulting changes in the size of holdings.[36]

With the implementation of land reform, the number of landholders increased because the previously neglected peasants who had no land in the pre-reform period were given land just as all other peasants of the area. But migration of members either to urban areas or from one peasant association to another is not possible. Individuals are left with only one alternative, to work on the land. The increasing rate of marriages also contributes to the demand for land. These factors cause pressure on the land and thereby fragmentation.[37] Under present circumstances, it would be more dangerous to carry out further redistribution of land as it would lead to further fragmentation. It has been observed in Wolmera district that there are some peasants who are willing to improve their land resources. But they are often held back by the fear that some portion of their land may be given away to the newly married people who wait for their share from the land redistribution process.[38] Since the continued fragmentation of land and scattered settlements have negative effects on agricultural production, it is more desirable to convince peasants about the advantages of co-operatives.

Agricultural production is also seriously hampered by the lack of improved inputs which depends upon the availability of cheap credit. The Government has taken various steps in this direction, but the amount of credit available is not enough for the peasants to fully use land and labour. Most of the Government budget for agriculture is directed to State Farms and very little is kept for small landowners. The establishment of more co-operatives will certainly help peasants in providing required credit at reasonable rates.

Price Policy

The farmer is generally in a weak bargaining position because of his need to sell produce immediately after harvest

36. *Ibid.*, p. 95.
37. *Ibid.*, p. 161.
38. *Ibid.*, p. 154.

in order to pay his debts. He also suffers from limited market information, uncontrolled weights and measures and the sheer efforts required for him to bring his produce to market over long distances. All these may force him to sell his produce even at low price. During the pre-revolutionary period, the government intervention in grain marketing had been weak. The Ethiopian Grain Board which was established with the aim of promoting price stability, suffered from shortcomings related to shortage of working capital and lack of a government price policy. Moreover, the EGB's share in the total wholesale grain trade was only 5 per cent and as such it could not exert any significant influence on prices.[39]

The need for the formation of prices through an improvement in the performance of the grain marketing system was evident to the Government after 1974. The Ministry of Agriculture had already some plans for the establishment of a new agency which could influence the prices of grains in the markets. The "Declaration on Economic Policy of Socialist Ethiopia" published in 1975 stated that "the Government may, where necessary, engage in wholesale and retail business in order to stabilise prices particularly of basic consumption items." In the grain marketing, the aims of the Government are to "assure stable producer and consumer prices ; maintain adequate producer incentives."[40] Accordingly, the Agricultural Marketing Corporation was established in order to exercise a major influence on prices.

The Government also adopted the outline of a new price policy for grains. While this policy did consider also the interests of consumers, it came out heavily in favour of the producers with a view to stimulating domestic production. The policy recognised that most of the marketable surplus of grains would be expected from the small holders composing the bulk of the population in the country. These small holders would have to be provided with adequate incentives to improve their production techniques to raise yield per hectare of land. In

---

39. John Holmberg, Grain Marketing and Land Reform in Ethiopia, *op. cit.*, p. 10.

40. *Ibid.*, p. 12.

the short period there would be no alternative to fertiliser for increasing the yields of small peasants. Other inputs like improved seeds, water resources, soil conservation measures, pesticides, etc., would come later.[41] It was suggested that the policy should take into account a ratio between the prices of fertilisers and grains that would provide incentives to the farmers for purchasing fertilisers.

But due to high cost of fertilisers it was not possible to follow that approach and ultimately it was decided to adopt other means than fertiliser. For example, the average producer price set by the Agricultural Marketing Corporation for different types of teff during the period between 1980 and 1983, as mentioned earlier, increased by 20 per cent in Dejen district, where as the price of fertiliser increased by more than 80 per cent.[42] Since the cost benefit approach did not win the required support, more emphasis has been paced on the consumers' interest than the producers.'

Peasants' Economic Situation

With the redistribution of land among the peasants, a new organisational structure has emerged and the peasant associations are becoming by increasingly involved in organising agricultural production. The peasants are now more alert and there is an increasing awareness that hard work can improve their economic position. But the available information based on various surveys and studies conducted so far clearly indicates that the peasants' economic situation has not improved much. According to one survey conducted in Western Wollega, the peasants of this area are almost as poor as they were before. The productive forces are as little developed as before the land reform. The study illustrates one concrete example in this respect.[43]

During the imperial regime Guddina was a tenant living in the countryside. He rented land from a rich peasant to which

41. *Ibid.*, p. 13.
42. Mengistu Woube, Problems of Land Reform Implementation in Rural Ethiopia, *op. cit.*, p. 104.
43. Michael Stahl, New Seeds in Old Soil, *op. cit.*, p. 70.

he paid 12 Birr per year for land use. But he could not use much of the land he rented, since he had no oxen. He could only afford to rent two oxen from a wealthier neighbour for less than a week annually. The total area ploughed was less than one hectare. He planted teff on the field and grew some pulses and coffee around the house. But the production was too insufficient for the family's subsistence. In order to survive Guddina worked as a wood-cutter and sold firewood to town people in a nearby market.

After land reform Proclamation he got cultivation rights to the plots of land he had rented from the rich peasant and all obligations to that man were abolished. Now he pays 4 Birr for the land use and 3 Birr as income tax. His total tax-burden is thus 7 Birr. He has now one ox which was given by the association on credit. He started to co-operate with a neighbour who had also got one ox. They worked out a system according to which each of them could use the two oxen on certain days a week. For the first time Guddina was able to cultivate an area of two hectares which would yield a harvest enough to release his family from semi-starvation.

Guddina obtained all means of production on credit. He got seeds from his association on credit. He also bought fertiliser from EPID on credit. Since he was not able to plough still another hectare of land due to lack of ploughing capacity and seeds, he is indebted. He owed the association 20 Birr for teff seeds and owed 12 Birr for fertiliser. Altogether his total debt amounted to 59 Birr to be paid before the next planting season (including oxen loan to be paid to the association). The study observes that on the basis of yield per hectare of land, Guddina was not able to produce enough to maintain his family. He would have to continue cutting wood for sale. The coffee he grew around the house would fetch approximately 15 Birr when sold on the market. This sum could cover the debt for the fertiliser he had got from EPID. But there remained a debt totalling 47 Birr to the peasant association and it seemed unlikely that he would be able to pay that sum. Guddina's situation is representative of a considerable portion of the peasantry. It means that the poor peasants at present do not produce any surplus which could be converted into cash and channelled into associations funds to finance collective investments.

The individual farms still continue to be in complete dominance. But they suffer from a number of problems. They still use traditional methods of cultivation and the consumption of inputs like fertilisers and improved seeds is very low. The crops are harvested in a traditional way without having sufficient storage facilities. The marketing and credit system does not provide any incentives to the peasants. Added to these are the problems of land fragmentation, absence of agricultural innovations and an inadequate infrastructural facilities. If the proper pre-conditions for improved agriculture were available, peasants could achieve higher output at lower production costs.[44]

The Ethiopian economy is still dominated by agriculture and that sector should be the principal focus of future development efforts. The development of agricultural sector will require a wide range of measures. The implementation of land reform did eliminate the major structural obstacle to its development. The rural population can now enjoy the fruits of the labour. But the mere redistribution of land and organisational changes cannot solve the basic agricultural problems. Land reform should not be seen in isolation from other major agricultural problems. Inequalities between households in the ownership of farm oxen and other implements and necessary inputs, as well as insufficient supply of agricultural credit, poor storage facilities, lack of all-weather roads and transport facilities[45] are some of the factors behind the present famine situation. Radical structural change can make sense only when it is integrated with a thoroughly worked out strategy for the overall agricultural development.

---

44. Mengistu Woube, Problems of Land Reform Implementation in Rural Ethiopia, *op. cit.*, p. 154.
45. *Ibid.*, p. 163.

Appendix I

Extracts from the Propaganda and Culture Committee, Ethiopia. A Decade of Revolution and Transformation, 1974-1984, Addis Ababa, 1984

Party Formation and Popular Participation

Party Formation

When the Revolution erupted in February, 1974 there was no organized political body to lead and co-ordinate the popular uprising of the masses. The PMAC, which assumed the leadership of the Revolution cognizant of the experiences of the various historical mass movements which failed owing to the absence of a well organized revolutionary leadership came to the conclusion that if the Ethiopian Revolution were to emerge victorious it had to have a Marxist-Leninist vanguard party as its foremost leader. And towards this end, the PMAC not only made a revolutionary call to the various revolutionary individuals and groups that had been working clandestinely to struggle for the purpose of organizing a vanguard party, but also created all the necessary conditions to hasten the process.

The actual process of forming the vanguard party was started with the establishment of the Provisional Office for Mass Organizational Affairs (POMOA), and the Yekatit 66 Political School in May 1976. The Revolutionary Government entrusted to these institutions the duty of propagating the science of Marxism-Leninism among the broad masses of Ethiopia and thus to create the appropriate pre-condition for the establishment of the party. As a result of the activities of these two political institutions and through the unreserved material and political assistance rendered by the Revolutionary Government five organizations were created by the Union of Ethiopian Marxist-Leninist Organizations in February 1977. These were the All Ethiopia Socialist Moverment (MEISONE) Revolutionary Struggle of the Oppressed of Ethiopia (ECHAT), Marxist-Leninist Revolutionary Organization (MALERID). Revolutionary Flame (SEDED), and Labour League (LEAGUE).

It was envisaged that in due course a merger among these five organizations which professed the acceptance of Marxism-Leninism would take place and the vanguard party of Ethiopia would be established. But since revolution involves contradictions, as soon as these organizations pronounced unity among themselves, problems and contradictions began to be manifested. In August 1977, just seven months after the formation of the Union, (MEISONE) abandoned the camp of the revolution on flimsy grounds in time of trying for the Revolutionary Motherland and joined the ranks of the counter-revolution (ECHAT), the other member of the union, owing to the erroneous political and ideological lines it had been advancing, was altogether expell from the Union in March, 1978.

These were not the only problems that beset the Union of Ethiopian Marxist-Leninist Organizations. As it is well known the principle of Democratic Centralism is the center piece for any Marxist-Leninist organization that aspires to expedite its historic mission correctly and efficiently. But in the case of the Union of Ethiopian Marxist-Leninist Organizations, each having its own organizational structure and programme, it was not possible to forge a common identity of views and to rally member organizations around one structure programme and leadership.

Faced with such difficulty the Revolutionary Government and genuine members of the different Marxist-Leninist organizations had to follow a new approach to the formation of the vangaurd party of Ethiopia. The new approach opted for was that instead of mechanically joining the various organizations, a center had to be established that could assemble all those genuine revolutionaries who had demonstrated their dedication and commitment to the cause of the masses in the course of the Revolution. The center that was to assemble all those genuine Marxists, was to be led by Comrade Mengistu Haile Mariam, who through his indefatigable work and commitment to the cause of the Revolution and through his wise and decisive leadership, and symbolized the determination of the Ethiopian masses to build a new socialist society free from all types of exploitation and oppression. And hence, the Commission for Organizing the Party of the Working People of Ethiopia (COPWE), which served as the new center for assembling the genuine revolutionaries of the country, came into being on December 18, 1979, with Comrade Mengistu Haile Mariam as its Chairman.

Although COPWE's misssion was to help expedite the process of party formation, it also at the same time acted as a vanguard party and as such, its organizational structure was very similar to that of any full-fledged Marxist-Leninist party. At the apex of its organizational structure, there was a Congress that formulated and approved all the relevant decisions that the Commission was to carry out in its tast of expediting the process of party formation. Next to the Congress, there was the Central Committee that guided the work of the Commission in between sessions of the Congress and next to the Central Committee, there was the Executive Committee that guided and oversaw the day to day activities of the Commission. In addition to thesc higher bodies, COPWE had also a net-work of organizational structures that connected it with the broad masses and were engaged in various aspects of social life. These are the regional, provincial and district committees as well as the primary organizations.

In its task of expediting the process of party formation COPWE was supposed to pass through three distinct phases. The first one was a phase of preparation which culminated in the calling of the First Congress of COPWE, which took place from June 16th to June 19th, 1980. During this phase all the necessary rules and regulations that govern the activities of the

Commission were prepared. These documents were presented to the First Congress of the Commission for approval.

The second phase was a phase of intense activities for laying down the organizational structures of the Commission. During this phase intensive recruitment of members was made, existing mass organizations such as the All-Ethiopia Trade Union, the All Ethiopia Peasants Association and the Urban Dwellers Associations were strengthened, and the Revolutionary Ethiopia Youth Association, the Revolutionary Ethiopia Woman's Association and the Working People's Control Committee were newly established on a nation-wide basis. In addition, during this period, the organizational structure of COPWE was extended to regions, sub-regions and districts with the primary organizations as the foundation. This phase culminated in the calling of the Second Congress of COPWE, which took place from January 4th to 7th, 1983.

The third and last phase concerned the further consolidation of COPWE both organizationally and ideologically and the final preparation for the formation of the vanguard party of Ethiopia. During this final phase, such important documents as party programme, party rules and regulations and other relevant documents were prepared for approval by the Founding Congress of the Workers Party of Ethiopia. This phase ended with the establishment of the Workers Party of Ethiopia in September, 1984. This was a truly historic occasion, not only to the Ethiopian revolutionaries who have struggled selflessly for almost a decade to build a truly vanguard party to guarantee the success of the Revolution, but also to the international revolutionary movement that struggles for the growth and consolidation of popular power and socialist construction everywhere.

Popular Participation

A popular revolution, above anything else, frees the broad masses from the political and socio-economic domination of the exploiting few ; it assures the supremacy of the masses over their former exploiters, guarantees their right and safeguards their vital interests. The cardinal objective set by the Ethiopian Revolution is stamp out the exploitation of man by man and the formation of the truly socialist society where every citizen would have a freedom, equality and fair share in the wealth of the country created by his labour.

In order to achieve these goals and guarantee the freedom, interests, rights and development of the broad masses, the key is revolutionary, organized and all-round participation of the masses themselves in all affairs that affect them and their country. The most fruitful participation is one which is purposeful, organized and consciously guided. This has been experienced by the Ethiopian masses during the last decade of revolutionary development which saw the flowering of democratic forums that embody popular power.

The All-Ethiopia Trade Union (AETU), representing the Ethiopian workers, the All-Ethiopia Peasants Association (AEPA), embracing the rural masses, the Urban Dwellers Association (UDA), Revolutionary Ethiopia Youth Association (REYA), Revolutionary Ethiopia Women's Association (REWA), the Professional Associations which have rallied the Ethiopian intelligentsia around the Revolution and the Working People's Control Committee are the main forums of popular participation in our country today.

Mass organizations and professional associations, embracing the active and creative workers, peasants, youth , women intellectuals, both in rural and urban areas and in all professions and sexes, have been acting as a natural bridge linking COPWE and the broad m asses and have become true custodians and representatives of the interests of the masses. These associations are tested schools of the masses training them in democratic experience class struggle and raising their consciousness. They are the solid basis and assurance for the realization of peoples' democracy and socialism in Ethiopia.

Each of the mass organization and professional association as a formidable forum in which popular interest and will are expressed and mass control is exercised, have clear objectives and tasks and elaborate organizational structures that are filled through active and conscious participation of the masses in the election and control of their leaders. Today, more than 1,300,000 persons are serving in the various elective bodies of the mass organizations.

The All-Ethiopia Trade Union was founded in 1977 and has a membership of over 300,00 strong and its foundation is a result of earlier struggle of many years. During the last decade the working class has grown considerably and scientific socialism has been disseminated among its ranks while the most conscious elements of it are being recruited, with growing emphasis, to the membership of the party. Through the Revolution therefore, conditions have become propitious for the working class to grow in number to mature ideologically and eventually to play its historical role as the most revolutionary class.

The working class, being guarded by COPWE (till WPEs formation) and AETU, its labour organization, is already playing a significant role to protect and promote the interest of its members and to develop and safeguard the revolutionary gains. This has been possible due to radical and democratic proclamations that have granted workers job security improvement of basic work conditions, job opportunity on equal basis, freedom of organization the possibility of participation in the management activities in their respective work places etc. This has generated enthusiasm in the workers to pay sacrifices in terms of raising productivity, in the efforts of national defence, literacy campaign expanding social services etc.

The workers as members of the society are also members of the UDA's, REYA, REWA and participate in the administrative and judicial work of the society.

The All-Ethiopia Trade Union links the Ethiopian workers with the struggle of the worldwide working people through the membership of international organizations and bilateral relations with organizations that struggle for the causes of peace, democracy and socialism. AETU is a member of the ILO, the Organization of African Trade Unions and has strong links with trade unions in the socialist and progressive countries.

The All-Ethiopia Peasant Association which embraces the peasantry, the most numerous social force of the Revolution, was founded in 1979 and consolidated through later actions. It has a membership of over five million, comprising both men and women who are grouped in 20,183 basic organizations.

The proclamation which made land public property was a

single most important act that freed the peasantry from feudal bondage and servitude. It also provided for the formation of peasants, association which ushered in an area of democratic participation in rural Ethiopia.

There are 458,774 peasants, men and women, who are elected to serve on various levels as members of the executive committees, judicial tribunals and control committees of their association.

AEPA has already proved to be powerful forum in enhancing, assisting and encouraging the peasants in the efforts of building new life in rural areas, spreading revolutionary ideas among the rural masses in order to raise their consciousness, and popularizing socialist production relation in order to eliminate the exploitation of man by man. AEPA has become a useful instrument to co-ordinate the participation of the peasants for the implementation of social projects, prepare the peasants for the safeguarding of the Revolution as well as the sovereignty unity and territorial integrity of the country ; establish and consolidate the unity of struggle between the peasantry and the other working people.

Thanks to the radical changes that have been effected in the rural areas, the peasants are no more an ignorant mass on which the idle and parasitic landlords and usurers flourish. They are riding themselves of illiteracy, ill-health and domination of any sort. Instead of being a docile object of exploitation, the peasants have become a conscious and organized power to reckon with. Guided by the invigorating ideals of the Revolution and led by COPWE (till WPE's formation) and AEPA, the Ethiopian peasants have already experienced new life and are daily registering new advances on the path towards building a just and prosperous society.

The idea of service co-operation spread in the rural areas from the early days of the Revolution. Over 3,500 service co-operatives have been formed and are rendering valuable assistance to the farmers in the task of marketing all essential goods in remote areas.

The ultimate aim of the Revolution is to help the peasants set up producers' co-operatives which will develop step by step towards higher levels This, of course, is to be done on a revolutionary basis and with full conviction and concern on the part of the peasants. There are over one thousand producers' cooperatives which are on various levels of development. Already, one can clearly see that the idea of cooperation is very popular among the rural population and the peasants have already seen that cooperative production is superior to the isolated and individualistic efforts that are known to them from time immemorial.

AEPA links the Ethiopian peasantry with similar movements all over the world in the struggle for peace, progress, democratic changes and socialism.

The Revolutionary Ethiopia Youth Associotion (*REYA*) was established in September, 1980. REYA is a culmination of the historic struggle of many years by the Ethiopian students to organize revolutionary youth under one umbrella in order to fight for a just and prosperous society. COPWE, which from the start attached great importance to the organization of the revolutionary youth, made all efforts for the establishment of REYA with its own constitution, a plan of action and internal rules and regulations.

REYA embraces youth between ages 14 to 30 and efforts are also being made to organize children between the ages of 7 to 13. Today, REYA has a membership of 3.8 million which is growing at a fast rate. There are 20,700 basic organizations spread in the breadth and width of the country.

The Association is run through elected leaders on every level. There are 453,000 elected office bearers in REYA's structure, the vast majority of whom give voluntary services. REYA organizes youth mainly on the basis of residential localitics. Some work is also being done to organize youth in their work places and educational establishments.

REYA's members have made laudable contributions in the construction of a new society. Their efforts in the economic field, particularly in rural agricultural work, in assisting in the settlement schemes, producers' cooperatives, clearing roads, building bridges, afforestation campaigns have been significant. Over the past three years, REYA members have planted 23 million seedlings and constructed around 70,000 kilometres long teraces to stop land erosion.

In the field of education, REYA members are the forefront contributors in the historic task of the literacy campaign which has earned Ethiopia fame and rewards in the international arena. 90% of the literacy teachers are youth.

Youth have also been actively contributing in expanding educational and health facilities to remote areas. Their role in the defence of the Revolution and the integrity of the country has been decisive. REYA has also been instrumental in spreading Marxism-Leninism among the young people.

REYA links the struggle of the Ethiopian youth with that of the world-wide youth movement both through membership in the international organizations such as IUS, WEDY, and All-Africa Students' Union and through bilateral contacts.

The founding, in September, 1980 of the Revolutionary Ethiopia Women's Association (REWA) was a result of the long struggle and untiring efforts of the Ethiopian progressive women to do away with oppression and exploitation in all its forms.

Both to liberate themselves from the archaic feudal prejudices and to contribute constructively towards the building of a new society, the Ethiopian women had to establish their own national organization.

Following the revolutionary directives, REWA strives to enable women to fully participate in the socio-economic and political efforts of the country and to help build new life for all. The objectives of the organization are set as : to disseminate scientific socialism among women ; to prepare women to occupy their appropriate position in society and to make active participation in productive social activities ; to make every effort to ensure that the right of women as mothers is recognized and they, as well as their children are well cared for ; to liberate women from political, economic and social dependence and prepare them to join hands with their class allies and fully participate in the struggle to build Socialist Ethiopia.

As a result of their own active participation in the emancipatory struggle which freed the masses from feudal bondage. The Ethiopian women have fully realized their worth and right and resolved never again to be oppressed and exploited. In this regard REWA's contribution is decisive.

REWA has mobilized over 5 million women in its membership. About 309,128 women participate in the running of the activities of REWA. These are elected members on various levels.

As one of the ardent fighters for the realization of people's democracy in Ethiopia REWA mobilizes the oppressed Ethiopian women to take part in multiple practical tasks both in urban and rural areas.

Through REWA's efforts women have made significant contribution in every respect to defend the gains of the Revolution and their motherland from attacks by internal and external enemies. Also laudable have been REWA's contribution in the fields of literacy, upgrading the skills of women, creating kindergartens, eradicating harmful practices and customs that undermine women, expanding various social services and recreation centres. REWA also publishes an informative and educative quarterly magazine in its efforts to raise the level of consciousness of the Ethiopian women.

In the international arena REWA has signed protocol agreements with women organizations of friendly countries and international organizations with the aim of promoting the struggle for peace, equality, national freedom, democracy, social progress, the rights of women and the welfare of children on international level.

With this in view, REWA has participated in many international conferences, arranged numerous successful national occasions and has made its contribution for peace and justice. REWA has been accepted as a bureau member of the Women's International Democratic Federation (WIDF).

The Urban Dwellers' Associations (*UDAs*) came into being with the historic proclamation of 1977 that nationalized all urban land and extra houses. However, UDAs as we know them now, developed with time through various decisive actions and became one of the formidable forces for the protection and growth of the Revolution.

UDAs mobilize the urban masses numbering about 4.4 million or about 15% of the total population of the country. The main aims of UDAs are :

Providing for urban dwellers to get organized in Kebele, Higher and Central Association and run their own affairs, solve their own problems, and directly participate in political, economic and social activities.

To enable the broad masses of urban dwellers to be armed with Marxism-Leninism and contribute their due shares in the struggle to liquidate from the land of Ethiopia feudalism, imperialism and bureaucratic capitalism.

As it is true with the other mass organizations, UDAs too are run by elected revolutionaries on various levels. The main elective bodies of UDAs are policy and public safety committees and judicial tribunals. The members are elected by the assemblies of the masses through secret ballot and must be trustworthy citizens committed to the cause of the masses and the Revolution.

The Kebele, which is the lowest unit in the structure of the UDAs is a very important democratic forum where all the city dwellers in their respective areas meet, discuss and decide upon all matters that affect them. Through their Kebeles, Higher and Central Organs, the broad urban masses engage in various political and socio-economic activities and strive to stamp out social evils and participate in the programmes of social defence, urban development, esiablishment of people's shops, cottage industries and other cooperatives. They also encourage urban dwellers to organize themselves and build their own houses according to accepted plan and to participate in the campaign to eradicate illiteracy. The current year is to witness the total cradication of illiteracy from the Ethiopian urban areas, a greatly significant result of the concerted struggle of the revolutionary people to do away with ignorance and backwardness.

In Ethiopia, intellectuals of various professions struggled for many years in order to organize themselves into professional associations and strove for the improvement of their profession and for the progress of the society. However, until the onset of the Revolution, all their attempts were confronted by stubborn opposition from the now defunct feudo-bourgeois regime which gave only lip service to such democratic practices.

Teachers, artists, writers, health workers and others strove in vain for the right of association and democratic change. But, today, as a result of the Revolution, people in various professions are establishing their own associations. To-date there are eight professional associations which are rallying thousands of intellectuals around the National Democratic Revolution Programme. These include teachers, medical professionals, veterinarians, journalists, writers, musicians, actors and painters. All the associations have pledged themselves to contribute unswervingly towards the improvement of their respective professsions and to struggle for the success of the Revolution, for world peace, justice, democracy and socialism.

Another important organ of mass participation, established with the view to ascertaining the control of the working people over the common wealth of the society against wasteful and parasitic individuals and counter-revolutionaries, is the Working People's Control Committee which came into being in 1981.

The main objectives of the Working People's Control Committee are :—

— to enable the masses to protect, conserve and administer through active participation, the resources and properties they have brought under their control,

— to work for the fulfilment of the National Democratic Revolution and to lay down the material foundation of socialism and to do away with the remnants of feudalism, imperialism and bureaucratic capitalism,

— to ensure that in all fields of engagement of the working people, the dignity of labour is properly appreciated, the quality and quantity of producion is augmented and discipline is promoted and respected,

— to follow up the proper protection of the properties of the Government and of organizations,

— to ensure the practical application of the principles : working with the participation of, and for, the people.

Follow-up and supervision of the proper implementation of proclamations, regulations and directives, and auditing and causing to be audited accounts of any government institution, organization, mass organization or private individuals are among some of the major tasks of the control committees.

The organizational structure of the WPCC follows the administrative structure of the country and runs from the national to regional, provincial, district and basic levels. The committees and their executives up to the district levels are elected from the forums of the mass organizations and those on basic levels are elected from the employees of each government institution, organization or cooperative society.

Though the WPCC is a new phenomenon in the Revolution, its members have already done a commendable job in terms of discharging their responsibilities. Already they have apprehended many idle, fraudulent and wasteful individuals and counter-revolutionaries of various shades thus protecting the property of the country and raising productivity.

In Ethiopia, popular participation or the process of giving power a deep social content, will gain new and solid ground under the leadership of the Workers' Party of Ethiopia which forms the core of the new political order. The party will hasten the process of building a new state and the establishment of the People's Democratic Republic of Ethiopia.

The new state apparatus will directly represent the working masses and reflect the interests of the entire working people, all nationalities and ethnic communities. As Comrade Mengistu Haile Mariam said on one occasion, "This state shall be one made up of representatives elected directly by the working people from amongst their own ranks and accountable to them. In short, it shall be a workers', peasants' and working peoples' state in general".

Further Developments

On 10 September 1984, Ethiopia announced the establishment of a Workers' Party (WP). Comrade Mengistu was named the party's Secretary-General, while an eleven-member Politburo was chosen and a 136-member Central Committee was elected by COPWE delegates. The Dergue was apparently left standing. Seven members of the Politburo and 20 per cent of the Central Committee are of the military. Ethiopia's official designation was altered to Ethiopian Socialist Democratic Republic. The creation of the Workers' Party suggests the elimination of COPWE and the completion of an important transitional phase of the Revolution.

APPENDIX II

Extracts from Negarit Gazeta, PROCLAMATION NO. 31 OF 1975. A PROCLAMATION TO PROVIDE FOR THE PUBLIC OWNERSHIP OF RURAL LANDS. (The extracts include the preamble, chapter 2 and chapter 3 of the proclamation).

"Ethiopia Tikdem"

Whereas, in countries like Ethiopia where the economy is agricultural a person's right, honour, status and standard of living is determined by his relation to the land ;

Whereas, several thousand *gashas* of land have been grabbed from the masses by an insignificant number of feudal lords and their families as a result of which the Ethiopian masses have been forced to live under conditions of serfdom ;

Whereas, it is essential to fundamentally alter the existing agrarian relations so that the Ethiopian peasant masses which have paid so much in sweat as in blood to maintain an extravagant feudal class may be liberated from age-old feudal oppression, injustice, poverty, and disease, and in order to lay the basis upon which all Ethiopians may henceforth live in eqaality, freedom, and fraternity ;

Whereas, the development of Ethiopia of the future can be assured not by permitting the exploitation of the many by the few as is now the case, but only by instituting the basic change in agrarian relations which would lay the basis upon which, through work by co-operation, the development of one becomes the development of all ;

Whereas, in order to increase agricultural production and to make the tiller the owner of the fruits of his labour, it is necessary to release the productive forces of the rural economy by liquidating the feudal system under which the nobility, aristocracy and a small number of other persons with adequate means of livelihood have prospered by the toil and sweat of the masses ;

Whereas, it is necessary to provide work for all rural people ;

Whereas, it is necessary to distribute land, increase rural income, and thereby lay the basis for the expansion of industry and the growth of the economy by providing for the participation of the peasantry in the national market ;

Whereas, it is essential to abolish the feudal system in order to release for industry the human labour suppressed within such system ;

Whereas, it is necessary to narrow the gap in rural wealth and income ;

Now, therefore, in accordance with Article 6 of the Definition of Powers of the Provisional Military Administration Council and its Chairman Proclamation No. 2/1974, it is hereby proclaimed as follows :

CHAPTER 2

PUBLIC OWNERSHIP OF RURAL LANDS

3. Public Ownership of Rural Lands

1. As of the effective date of this Proclamation, all rural lands shall be the collective property of the Ethiopian people.
2. No person or business organization or any other organization shall hold rural land in private ownership.

3. No compensation shall be paid in respect of rural lands and any forests and tree-crops thereon ; provided that fair compensation shall be paid for movable properties and permanent works on the land.

4. Distribution of Land to the Tiller in Provinces with Privately Owned Rural Lands

1. Without differentiation of the sexes, any person who is willing to personally cultivate land shall be allotted rural land sufficient for his maintenance and that of his family.
2. Any person who has been a landowner and is willing to personally cultivate land shall likewise be allotted land.
3. The size of land to be allotted to any farming family shall at no time exceed 10 hectares (1/4 of a gasha).
4. The size of land to be allotted to farming families shall as far as possible be equal provided that it may vary depending on the local conditions and the productive potential of the land.
5. No person may use hired labour to cultivate his holding ; provided that the foregoing prohibition shall not apply to a woman with no other adequate means of livelihood, or where the holder dies, is sick or old, to the wife or the husband or to his or her children who have not attained majority :
6. The Minister shall determine by regulations the size of land sufficient for the maintenance of a farming family.

5. Prohibition of Transfer of Land

No person may by sale, exchange, succession, mortgage, antichresis, lease or otherwise transfer his holding to another ; provided that upon the death of the holder, the wife or husband or minor children of the deceased or where these are not present, any child of the deceased who has attained majority, shall have the right to use the land.

6. Tenant's Holding

1. Until all lands are distributed pursuant to Chapter 3 of

this Proclamation, any tenant or hired labourer shall have possessory right over the land he tills ; provided that a resident landowner who has leased out all his lands shall have the right to equally share the land with his tenants. The Government shall protect the rights of such landowner in any way it thinks fit.

2. The provisions of the preceding sub-article shall not apply to a person who has rented land from a woman with no other adequate means of livelihood or from a person who by reason of illness, old age or youth, cannot personally cultivate his holding.
3. As of the effective date of this proclamation the relationship between landowner and tenant is abolished. Accordingly, the tenant shall be free from payment to the landowner of rent, debts or any other obligation. Likewise, a landowner who has given his land in antichresis to a tenant shall be free from the payment of his debt.
4. The tenant shall have the right to retain agricultural implements and a pair of farm oxen belonging to the landowner for which a reasonable compensation shall be paid to the landowner within a period not exceeding three years ; provided that a landowner with no other agricultural implements and farm oxen or no adequate means of livelihood may take back his implements and oxen from the tenant.

7. Large-scale Farms

1. As of the effective date of this proclamation, any large-scale farm shall be organized as a State or a co-operative farm or shall be allotted to tillers ; provided that until the establishment of State or co-operative farms the Government shall administer such farms in any manner it deems fit and provided further that until the Government decides upon the manner in which such farms are to be administered, the ex-owners of these farms shall have the responsibility to continue administering them.
2. The Government shall pay fair compensation for mov-

able property and permanent works on such farms ; provided that compensation shall not be paid for the value of the land.

CHAPTER 3

Establishment of Associations for the Implementation of the Proclamation

8. Establishment of Peasant Associations

To carry out the provisions of this proclamation a peasant association shall be formed, in a manner suitable for development purposes, taking a *chika* area as a basis, within a minimum area of 800 hectares (20 gashas).

9. Members of the Association

1. A tenant, a landless person, a hired agricultural worker of land-owner with less than 10 hectarcs of land may become a member of the Association.
2. After completion of the distribution of land pursuant to Article 10 (1) hereof, any person who had been a land-owner may become a member of the Association.

10. Functions of Peasant Associations

The functions of peasant associations shall be the following :

1. to distribute, with the solicited assistance of the Government when necessary, land forming the area mentioned in Article 8 as much as possible equally, and in the following order :

   (*a*) to farmer tenants and former landowners residing within the area,
   (*b*) to evicted tenants,
   (*c*) to persons who reside within the area but do not have work or sufficient means of livelihood,
   (*d*) to farmers coming from outside of the area,
   (*e*) to pensioned persons who are willing to undertake personal cultivation,

(*f*) to organizations needing land for their upkeep ;

2. to follow land-use directives to be issued by the Government ;
3. to administer and conserve any public property within the area especially the soil, water and forest ;
4. to establish judicial tribunals to hear land disputes arising within the area ;
5. to establish marketing and credit co-operatives and other associations like the *debo* which would help farmers to co-operate in manual and other works ;
6. to build with the co-operation of the Government schools, clinics and similar institutions necessary for the area ;
7. to cultivate the holdings of persons who, by reason of old age, youth or illness, or in the case of a woman, by reason of her husband's death, cannot cultivate their holdings ;
8. to undertake villagization programmes ;
9. to exclude from distribution mining and forest lands and places of historical and antiquarian significance.

Ethiopia : Statistical Tables

TABLE I : Area and Population

AREA, POPULATION AND DENSITY

Area (sq km)	1,223,600
Population (official estimates at mid-year)	
1979	30,251,000
1980	31,065,000
1982	32,775,000
Density (per sq km) 1982	26.8

*472,435 sq miles.

ADMINISTRATIVE REGIONS (July 1980 estimates)

	Area (sq km)	Population
Arussi	24,600	1,149,400
Bale	128,300	879,200
Eritrea	177,400	2,426,200
Gemu Goffa	40,100	1,003,400
Gojam	64,500	2,037,900
Gondar	73,400	2,053,400
Hararge	254,800	3,125,220
Illubabor	50,800	810,800
Kefa (Kaffa)	53,000	1,615,400
Shoa*	85,500	6,362,200
Sidamo	116,700	2,808,300
Tigre	65,700	2,162,100
Wollega	69,800	2,019,200
Wollo	79,000	2,612,600
Total	1,223,600	31,065,300

* Includes the capital, Addis Ababa, which is also a separate Administrative Region.

(*Contd.*)

PRINCIPAL TOWNS (estimated population at July 1980)

Addis Ababa (capital)	1,277 159	Harar	62,921
Asmara	424,532	Bahr Dar	52,188
Dire Dawa	82,024	Debre Zeit	49,570
Gondar (incl. Azeso)	76,932	Mekele	46,846
Dessie	76,616	Debre Markos	40,686
Nazret	69,865	Keren	37,931
Jimma	63,837	Akaki	35,022
		Assela	34,874
		Massawa	32,977

BIRTHS AND DEATHS

Average annual birth rate 49.8 per 1,000 in 1970-75 and 1975-80 ; death rate 25.4 per 1,000 in 1970-75, 25.2 per 1,000 in 1975-80 (UN estimates).

ECONOMICALLY ACTIVE POPULATION

(ILO estimates, '000 persons at mid-1970)

	Males	Females	Total
Agriculture, etc.	5,866	3,240	9,106
Industry	394	233	627
Services	782	307	1,090
Total	7,042	3,781	10,823

Source : ILO, *Labour Force Estimates and Projections, 1950-2000.*

Mid-1981 (estimates in '000) : Agriculture, etc. 10,350 ; Total 13,176.

Mid-1982 (estimates in '000) : Agriculture, etc. 10,457 ; Total 13,413.

(Source : FAO, *Production Yearbook*).

TABLE 2 : **Temperatures and Precipitation**

Absolute altitudes and latitudes	Average temperature, °C, and precipitation, mm	Jan.	Feb.	March	Apr.	May	June	July	Aug.	Sep.	Oct.	Nov.	Dec.	Average annual values	Number of days with precipitation a year
Addis Ababa 2,408 m 9°02′	D*	23	24	24	24	24	23	20	20	21	22	22	22	22	
	N**	7	9	9	11	11	10	10	10	10	8	7	7	9	
	P***	*12*	*36*	*76*	*99*	*70*	*127*	*264*	*296*	*186*	*33*	*13*	*10*	*1,220*	141
Aseb 10 m 13°01′	D	31	32	34	36	37	37	41	37	37	37	34	31	36	
	N	22	21	23	26	27	29	31	30	28	26	24	20	25	...
	P	*7*	*8*	*0*	*0*	*0*	*0*	*0*	*0*	2	*0*	*1*	*1*	*19*	
Asmara 2,325 m 15°17′	D	22	24	25	25	25	25	21	21	23	22	21	21	23	
	N	6	7	9	10	11	12	11	12	10	9	8	6	9	
	P	*1*	3	*15*	*23*	*47*	*28*	*166*	*173*	*21*	*16*	*10*	*0*	*503*	56
Bahr Dar 1,790 m 11°36′	D	26	27	29	29	28	26	24	23	26	25	25	25	26	
	N	6	7	10	11	13	13	14	14	13	12	11	7	11	
	P	*1*	*0*	*10*	*28*	*76*	*114*	*422*	*292*	*231*	*90*	*27*	*0*	*1,291*	115
Bonga 1,650 m 7°15′	D	28	28	28	28	27	26	26	25	25	26	27	28	27	
	N	10	11	18	11	12	11	12	12	12	11	10	9	11	
	P	*69*	*101*	141	*178*	*266*	*235*	*177*	*151*	*126*	*131*	*32*	*43*	*1,650*	...

(*Contd.*)

Absolute altitudes and latitudes	Average temperature, °C, and precipitation, mm	Jan.	Feb.	March	Apr.	May	June	July	Aug.	Sept.	Oct.	Nov.	Dec.	Average annual values	Number of days with precipitation a year
Debre Markos	D	23	24	25	25	24	21	19	19	20	21	22	22	22	
2,509 m	N	7	8	9	10	10	9	10	10	9	8	8	6	9	
10°22′	*P*	*11*	*23*	*48*	*59*	*77*	178	*309*	*289*	*209*	*53*	*49*	*25*	*1,330*	...
Dire Dawa	D	28	29	32	31	33	34	32	31	32	32	30	27	31	
1,160 m	N	13	16	17	19	21	21	20	19	19	18	16	14	18	
9°36′	*P*	2	*25*	*60*	*77*	*70*	*30*	*110*	*120*	*70*	*10*	*9*	*16*	*602*	76
Gambela	D	37	38	39	37	34	32	31	31	32	33	35	36	35	
410 m	N	18	20	21	22	21	21	20	20	20	20	19	18	20	
8°14′	*P*	*6*	*9*	*35*	*86*	*154*	*164*	*219*	*243*	*181*	*94*	*47*	*13*	*1,251*	103
Goba	D	19	19	19	18	19	20	20	20	19	18	16	18	19	
2,500 m	N	4	5	6	8	8	7	7	8	7	7	6	4	7	
7°00′	*P*	*32*	*35*	*35*	*165*	*104*	*53*	105	150	*119*	*100*	*86*	*15*	*1,003*	...
Gondar	D	27	28	28	28	27	23	20	21	23	25	26	26	25	
2,120 m	N	10	11	12	13	13	10	9	9	10	10	10	10	11	
12°36′	*P*	*1*	*3*	*9*	*58*	*80*	*199*	*364*	*381*	*107*	*59*	*6*	*3*	*1,276*	118
Gore	D	24	25	26	25	23	21	20	20	21	22	22	23	23	
2,002 m	N	14	14	15	14	14	13	13	12	12	13	13	13	13	
8°10′	*P*	*39*	*46*	*83*	153	*267*	*328*	*310*	*360*	*330*	*176*	*110*	*43*	*2,245*	180

(*Contd.*)

Absolute altitudes and latitudes	Average temperature, °C, and precipitation, mm	Jan.	Feb.	March	Apr.	May	June	July	Aug.	Sep.	Oct.	Nov.	Dec.	Average annual values	Number of days with precipitation a year
Jimma	D	28	28	28	28	27	25	23	24	25	26	26	27	26	
1,740 m	N	7	10	12	13	13	13	13	13	14	10	10	8	11	
7°39′	*P*	*17*	*55*	*115*	*126*	*164*	*228*	*232*	*222*	*188*	*95*	*80*	*18*	*1,540*	207
Massawa	D	30	29	31	33	36	40	41	40	39	36	33	31	35	
5 m	N	22	22	24	16	27	28	31	30	29	26	23	23	26	
15°36′	*P*	*39*	*12*	*9*	*10*	*8*	*0*	*0*	*10*	*2*	*2*	*46*	*21*	*160*	...
Neghele	D	31	32	29	27	26	25	24	24	28	26	28	29	28	
1,444 m	N	15	16	17	15	15	13	13	13	13	13	13	13	14	
5°19′	*P*	*5*	*28*	*118*	*199*	*184*	*2*	*9*	*1*	*31*	*118*	*68*	*3*	*786*	80
Nekemte	D	25	26	27	28	26	24	20	19	22	22	25	27	24	
2,000 m	N	13	13	14	14	13	12	12	12	12	12	12	11	12	
9°05	*P*	*19*	*42*	*61*	*82*	*169*	*353*	*321*	*365*	*258*	*90*	*31*	*7*	*1,798*	—
Teseney	D	34	36	38	40	41	39	35	38	35	38	26	34	36	
615 m	N	17	18	21	23	25	24	23	22	22	22	22	19	21	
15°07′	*P*	0	*0*	*0*	*0*	*16*	*12*	*76*	*132*	44	*13*	*17*	*0*	*310*	—

* Diurnal Temperature

** Nocturnal

*** Precipitation

TABLE III : **Gross Domestic Product by Economic Activity**

(million Birr at current prices, year ending 7 July)

	1979/80	1980/81	1981/82
Agriculture and livestock	3,723.6	3,927.1	3,879.1
Forestry	143.6	147.1	151.0
Hunting	1.3	1.3	
Fishing	3.6	3.6	4.9
Mining and quarrying	8.1	8.7	9.8
Manufacturing	533.5	561.1	600.1
Handicraft and small industry	296.6	307.0	317.8
Building and construction	301.0	395.0	325.9
Electricity and water	54.5	64.9	59.5
Wholesale and retail trade	812.2	870.5	889.2
Transport and communications	355.1	383.5	434.0
Banking, insurance and real estate	244.5	261.1	304.4
Public administration and defence	510.2	552.0	617.4
Ownership of dwellings	200.2	206.3	212.0
Educational services	164.0	179.4	
Medical and health services	48.2	52.4	319.0
Domestic services	67.3	68.0	
Other	161.7	167.4	174.1
GDP at factor cost	7,629.0	8,156.4	8,298.2
Indirect taxes, *less* subsidies	800.9	949.1	820.6
GDP at market prices	8,429.9	9,105.5	9,118.8

Source : National Revolutionary Development Campaign and Central Planning Supreme Council Secretariat.

TABLE-IV : Main Agricultural Crops

('000 metric tons)

	1980	1981	1982*
Wheat	522	1614	650
Barley	1,075	1,150*	1,150
Maize	948	1,100*	1,000
Oats	46	12*	12
Millet (Dagusa)	205	200*	200
Sorghum	1,642	1,411	1,300
Other cereals	1,430	1,312	1,400
Potatoes*	238	240	240
Other roots and tubers*	1,176	1,228	1,280
Dry beans	20	20*	22
Dry peas	148	200*	190
Dry broad beans	469	450*	450
Chick-peas	147	118*	150
Lentils	31	60	60
Other pulses	100	130	130
Sugar cane	1,404	1,380*	1,480
Groundnuts (in shell)	20†	20†	20*
Castor beans	12	12	12
Rape seed*	22	22	22
Linseed	27	28*	28*
Sesame seed	36†	35*	36*
Seed cotton†	89	89	89
Cottonseed†	59	59	59
Cotton (lint)†	27	27	27
Safflower seed*	65	65	66
Coffee (green)	187†	202	202
Vegetables and melons*	484	496	499
Bananas*	73	73	73
Other fruit (excl. melons)*	127	130	133
Tobacco (leaves)*	3	3	3
Fibre crops (excl. cotton)*	15	15	15

*FAO estimates. †Unofficial estimates.

Source : FAO, *Production Yearbook*

TABLE V : **Coffee Inspected by the Coffee and Tea Development Marketing Corporation at Inspection Stations**

(Tons)

Districts	1979-80	1980-81
Arussi	N.A.	N.A.
GamoGofa	2895	1584
Illubabor	7187	11805
Kaffa	27148	26862
Shewa	3664	2163
Sidamo	37700	31084
Wellega	12511	17693

Source : CSO, Ethiopia—Statistical Abstract, 1980, Addis Ababa, 1982.

TABLE VI : **Number of Livestock and Livestock Products**

Number of Livestock ('000 head)

	1980	1981	1982
Cattle	26,000	26,100	26,200
Sheep	23,250	23,300	23,350
Goats	17,180	17,200	17,220
Asses	3,890	3,895	3,900
Horses	1,530	1,540	1,557
Mules	1,445	1,450	1,400
Camels	980	990	1,000
Pigs	18	18	19
Poultry	53,000	53,500	54,000

Source : FAO, *Production Yearbook*.

Livestock Products ('000 metric tons) *(Contd.)*

	1980	1981	1982
Beef and veal	214	214	215
Mutton and lamb	77	77	77
Goats' meat	55	55	55
Pig meat	1	1	1
Poultry meat	59	60	60
Other meat	61	64	67
Edible offals	85	86	86
Cows' milk	590	600	610
Goats' milk	95	95	95
Sheep's milk	58	58	58
Butter	8.9	8.9	9.0
Hen eggs	73.1	73.8	74.5
Honey	20.5*	20.7	20.8
Wool :			
greasy	12.2	12.3	12.3
clean	6.3	6.3	6.3
Cattle hides	41.0	41.1	41.3
Sheep skins	13.8	13.8	13.8
Goat skins	11.7	11.7	11.7

*Unofficial estimate.

Source : *FAO*, *Production Yearbook* and *Monthly Bulletin of Statistics.*

TABLE VII : **Principal Industrial Products**

		1977-78	1978-79	1979-80
Tinned meat	metric tons	108	64	25
Frozen meat	metric tons	362	219	1,175
Wheat flour	metric tons	80,963	114,824	190,536
Macaroni	metric tons	6,569	14,091	53,609
Refined sugar	metric tons	159,759	155,336	153,504
Salt (sea)	metric tons	40,119	107,433	121,977
Pasteurized milk	hectolitres	72,058	79,921	112,037
Liqueurs	hectolitres	60,445	30,217	33,133
Wine	hectolitres	65,980	71,044	76,710
Beer	hectolitres	414,976	495,521	552,052
Lemonade	hectolitres	327,872	264,333	710,813
Mineral waters	hectolitres	144,876	1,139,264	160,182
Cigarettes	'000	1,004,691	1,405,606	1,458,093
Cotton yarn	metric tons	7,882	7,851	9,259
Woven cotton fabrics	'000 sq metres	65,460	88,140	89,266
Woollen blankets	number	457,110	300,700	549,337
Wollen carpets	sq metres	5,904	15,600	20,504
Nylon fabrics	'000 sq metres	5,148	5,911	5,906
Hosieries	dozen	196,647	400,494	528,032
Leather footwear	pairs	1,038,899	1,491,416	1,498,721
Canvas and rubber footwear	pairs	916,715	2,313,519	2.965,041
Plastic footwear	pairs	561,401	1,709,078	2,191,209
Timber	cubic metres	66,015	41,105	39,183
Plywood	cubic metres	2,142	2,418	2,418
Particle board	cubic metres	7,568	2,164	2,173
Soap	metric tons	8,951	8,777	10,296
Ethyl alcohol	hectolitres	10,599	5,995	7,409
Liquefied petroleum gas	metric tons	2,982	4,131	n-a.
Gasoil	metric tons	175,740	206,323	145,039
Regular benzine	metric tons	70,438	88,092	75,144
Gasoline	metric tons	5,077	n.a.	n.a.
Fuel oil	metric tons	232,311	298,285	308,020
Diesel oil	metric tons	8,742	1,154	4,788
Oxygen	cubic metres	114,646	139,683	149,378
Matches	'000 boxes	18,493	16,220	34,989
Clay building bricks	'000	13,666	19,884	22,825
Quicklime	metric tons	3,491	6,796	8,848
Cement	metric tons	103,480	95,409	123,365
Nails	metric tons	2,784	3,693	4,913
Electric enegry	'000 kwh	452,464	543,782	n.a.

Source : Europa Publication, Africa South of the Sahara 1984-85, Europe Publication Ltd. 1984.

TABLE VIII : External Trade

Principal Commodities ('000 Birr)

Imports c.i.f.	1979	1980	1982
Food and live animals	61,114	92,358	124,454
Beverages and tobacco	14,578	11,875	20,265
Crude petroleum	196,066	262,451	n.a.
Petroleum products	33,659	107,091	389,383
Chemicals	122,103	167,505	110,102
Medicinal and pharmaceutical products	36,620	39,969	38,457
Soaps, cleaning and polishing preparations	13,070	17,727	24,540
Rubber products	15,073	36,281	24,837
Papei and paper manufactures	16,998	36,114	17,808
Textile yarn, fabrics, etc.	50,186	40,765	44,686
Clothing	10,874	10,561	14,130
Metals and metal manufactures	89,469	107,366	156,426
Machinery (incl. aircaft)	199,950	174,413	238,710
Electrical machinery, apparatus, etc.	59,998	53,835	80,856
Road motor vehicles	137,090	151,592	190,498
Total (incl. others)	1,175,053	1,494,703	1,528,829

1981 : Total imports 1,526.6 million Birr.

(*Contd.*)

Exports f.o.b.*	1979	1980	1982
Live animals	4,143	10,936	11,457
Meat, canned and frozen	3,581	6,316	4,701
Pulses	22,472	23,802	36,568
Fresh fruit	1,113	1,970	4,893
Vegetables, fresh and frozen	2,984	4,801	
Sugar	10,689	9,157	9,623
Coffee	592,647	563,649	511,538
Spices	3,327	3,976	1,776
Oilseed cake	10,796	4,743	5,862
Hides and skins	141,337	107,947	84,538
Oilseeds	9,270	15,492	24,749
Civet	1,348	1,104	835
Incense	1,358	148	3,346
Chat	10,508	24,732	32,858
Beeswax	3,858	4,192	1,757
Total (incl. others)	863,608	878,725	835,441

*Excluding re-exports ('000 birr) : 717 in 1979 ; 25 in 1980.

Source : National Bank of Ethiopia, *Quarterly Bulletin.*
1981 (million birr) : Coffee 469.9 ; Hides and skins 98.4 ; Pulses 25.1 ; Total (incl. others) 782.7 (*Source* : IMF, *International Financial Statistics*).

TABLE IX : **Principal Trading Partners ('000 Birr)**

Imports	1979	1980	1982
China, People's Rep.	13,355	20,742	8,210
France	19,486	32,634	52,111
German Dem. Rep.	36,752	34,666	31,848
Germany, Fed. Rep.	118,585	142,387	157,877
India	14,111	13,156	n.a.
Israel	15,925	21,317	n.a.
Italy	117,244	161,477	185,504
Japan	124,399	129,715	145,431
Kenya	15,779	15,596	n.a.
Korea, Rep.	22,347	14,557	n.a.
Kuwait	181,515	69,225	n.a.
Netherlands	23,870	33,986	30,424
Saudi Arabia	1,386	1,811	17,957
Sweden	36,741	22,249	n.a.
Switzerland	22,068	52,736	n.a.
Taiwan	13,734	11,494	n.a.
USSR	22,857	287,746	441,821
United Kingdom	77,871	97,930	122,810
USA	137,492	114,885	63,568
Total (incl. others)	1,175,053	1,494,703	1,652,829

*Figures for 1981 are not available.

(*Contd.*)

Exports	1979	1980	1982
Belgium	10,946	10,506	n.a,
Djibouti	50,923	100,351	77,519
France	30,342	58,399	44,311
German Dem. Rep.	764	3,925	15,422
Germany, Fed. Rep.	6I,969	74,593	133,571
Israel	3,330	3,368	n.a.
Italy	97,820	89,150	43,913
Japan	53,419	56.645	62,172
Netherlands	32,007	17,884	10,504
Saudi Arabia	75,744	66,420	56,015
Spain	12,932	7,680	n.a.
Switzerland	3,325	1,107	n.a.
USSR	58,353	81,193	5,042
United Kingdom	19,503	18,451	16,927
USA	246,134	59,621	216,045
Yemen, People's Dem. Rep.	11,374	11,437	n.a.
Yugoslavia	34,469	34,169	15,154
Total (incl. others)	864,327	878,750	836,295

Source : National Bank of Ethiopia, *Quarterly Bulletin.*

TABLE X : **Balance of Payments** (US $ million)

	1980	1981	1982
Merchandise exports f.o.b.	419.3	374.1	402.8
Merchandise imports f.o.b.	−649.6	−629.8	−675.2
Trade Balance	−230.3	−255.7	−272.4
Exports of Services	153.5	157.1	176.5
Imports of Services	−221.8	−219.9	−239.8
Balance on goods and services	−298.6	−318.5	−335.7
Private unrequired transfers (net)	18.2	23.7	38.1
Government unrequired transfers (net)	53.7	44.8	102.4
Current balance	−226.7	−250.0	−195.2
Long-term capital (net)	59.2	242.4	69.9
Short-term capital (net)	99.1	123.4	39.5
Net errors and omissions	−35.2	—7.6	10.4
Total (net monetary movements)	−103.7	108.3	−75.4
Monetization of gold	12.6	—	—
Allocation of IMF special drawing rights	5.0	4.7	—
Valuation changes (net)	—22.3	18.5	−0.6
IMF trust fund loans	10.8	0.1	—
Changes in reserves	−97.6	131.6	−76.0

Source : IMF, *International Financial Statistics.*

TABLE XI

Budget Estimates (million Birr, year ending 7 July)*

Revenue	1979-80	1980-81	1983-84†
Direct taxes	361.4	381.2	638.2
Domestic indirect taxes	342.7	384.8	495.5
Taxes on foreign trade	623.3	578.5	572.1
Charges and fees	11.8	22.4	23.9
Sales of goods and services	32.1	37.7	43.0
Property and investment	137.0	224.4	439.1
Miscellaneous	21.0	6.0	7.0
Pension contributions	17.9	19.9	26.0
External assistance	151.6	216.8	219.4
Capital receipts	273.0	312.5	433.4
Total	1,971.8	2,184.2	2,897.6

Expenditure	1979-80	1980-81	1983-84†
Current			
National defence	750.8	736.8	869.6
Internal order and justice	143.7	150.6	174.0
Organs of state	104.3	113.4	161.7
Public works and communications	43.0	44.8	63.2
Agriculture, industry, commerce and mining	75.8	86.7	96.9
Education and culture	221.2	229.7	252.0
Public health and social welfare	114.6	130.4	136.2
Pensions	66.0	82.6	100.0
Public debt	92.0	107.8	192.6
Bank charges	1.5	2.0	2.8
Unallocated	41.6	56.3	107.6
Total current	1,654.6	1,741.1	2,224.8
Capital			
Economic development	594.6	795.5	1,078.2
Social development	89.1	120.6	148.2
General services	26.7	21.2	8.6
Total capital	710.4	937.3	1,234.9
Total	2,365.0	2,678.4	3,459.8

*Figures for 1981-82 and 1982-83 are not available.

*Projected.

Ethiopian Position in Sub-Saharan Africa

Table I : Basic Indicators

	Population (millions) mid-1984	Area (thousands of square kilometres)	GNP per capita Dollars 1984	GNP per capita Annual growth rate 1965-84 (percent)	Annual rate of inflation (percent) 1965-73	Annual rate of inflation (percent) 1973-83
Low-income economics	260.6t	15,694t	220w	(.)w	3.6w	18.5w
1 Ethiopia	42.9	1,222	110	0.5	1.8	4.4
2 Mali	7.3	1,240	140	1.2	7.6	10.3
3 Zaire	30.6	2,345	140	—1.3	18.7	48.2
4 Burkina	6.6	274	160	1.4	2.6	10.8
5 Guinea-Bissau	0.9	36	180	—	—	6.9
6 Niger	6.3	1,267	190	—1.2	4.0	11.8
7 Malawi	6.8	118	210	2.2	4.5	9.8
8 Tanzania	21.5	945	210	0.9	3.2	11.5
9 Burundi	4.6	28	220	2.1	2.9	12.4

10 Uganda	14.3	236	230	—4.4a	5.6	62.7b
11 Togo	2.9	57	250	1.1	3.1	8.3
12 Gambia	0.7	11	260	1.4	3.0	10.4
13 Somalia	5.2	638	260	—0.8a	3.8	20.1
14 Benin	3.9	113	270	1.0	3.6	10.8
15 Central African Rep.	2.5	623	270	0.1	3.0	14.4
16 Madagascar	9.7	587	270	—1.2	4.1	13.9
17 Rwanda	5.9	26	270	2.3	7.7	11.2
18 Guinea	5.9	246	300	1.1	3.0	4.0
19 Kenya	19.7	583	300	2.3	2.3	10.8
20 Sierra Leone	3.7	72	300	1.1	1.9	14.7
21 Sudan	21.5	2,506	340	1.3	7.2	18.0
22 Ghana	13.4	239	350	—2.1	8.1	51.6
23 Senegal	6.4	196	380	—0.5	3.0	8.9
24 Chad	4.9	1,284	—	—	4.5	8.3b
25 Mozambique	13.4	802	—	—	—	—
Middle-income oil importers	32.6t	3,257t	620w	1.2w	3.5w	10.8w
26 Mauritania	1.7	1,031	450	0.3	3.9	7.8
27 Liberia	2.1	111	470	0.8	1.5	7.2
28 Zambia	6.5	753	470	--1.3	5.2	10.3

(Contd.)

29 Lesotho	1.5	30	530	6.3	4.4	11.9
30 Cote d' Ivoire	9.9	322	610	1.0	4.4	11.9
31 Zimbabwe	8.2	391	740	1.5	3.0	9.7
32 Swaziland	0.7	17	800	2.6	4.3	14.1
33 Botswana	1.0	600	910	8.5	4.4	9.8
34 Mauritius	1.0	2	1,100	2.8	5.6	13.1
Middle-income oil exporters	117.7t	3,256t	800w	3.4w	8.3w	13.2w
35 Nigeria	96.8	924	770	3.2	10.3	13.3
36 Cameroon	9.9	475	810	2.7	5.8	12.6
37 Congo, People's Rep.	1.8	342	1,120	3.5	4.6	12.4
38 Gabon	0.8	268	3,480	3.2	5.8	18.5
39 Angola	8.4	1,247	—	—	—	—
Sub-Saharan Africa	410.9t	22,207t	420w	1.8w	4.2w	14.8w

The letters, w, m, t. indicate weight averages, medians and totals respectively

a. Because data for the entire period are not available, figures are for periods other than that specified.

b. Figures are for 1973-82, not 1973-83.

c. Figures are for 1983, not 1984.

d. Figures are for 1965-83, not 1965-84

*Tables I to VI are based on the Report of The World Bank, Financing Adjustment with Growth in Sub-Saharan Africa, 1986-90, Washington, 1986.

TABLE II : Structure of Production

	GDP (millions of dollars)		Distribution of Gross Domestic Product (percent)					
			Agriculture		Industry		Services	
	1965	1983	1965	1983	1965	1983	1965	1983
Low-income economies			43w	40w	16w	17w	41w	43w
1 Ethiopia	1,180	4,270	58	48	14	16	28	36
2 Mali	370	980	49	46	13	11	38	43
3 Zaire	1,640	5,440d	22	36	27	20	51	44
4 Burkina	250	900	52	41	15	19	32	40
5 Guinea-Bissau	—	145	—	43	—	11	—	46
6 Niger	370	1,340	63	33	9	31	28	37
7 Malawi	220	1,330	50	—	13	—	37	—
8 Tanzania	790	4,550	46	52	14	15	40	33
9 Burundi	160	1,020	—	58	—	16	—	26
10 Uganda	1,080	3,360d	52	—	13	—	35	—
11 Togo	190	720	45	22	21	28	34	50
12 Gambia	40	220	39	27	10	14	50	59

(*Contd.*)

13 Somalia	220	1,540	71	50	6	11	24	39
14 Benin	210	930	53	40	9	14	24	47
15 Central African Rep.	140	600	46	37	16	21	38	42
16 Madagascar c, d	730	2,850	31	41	16	15	53	44
17 Rwanda	150	1,560	75	—	7	—	18	—
18 Guinea	520	1,910	—	38	—	23	—	39
19 Kenya	920	4,940	35	33	18	20	47	46
20 Sierra Leone	320	950	34	32	28	20	38	48
21 Sudan	1,330	6,850	54	34	9	15	37	51
22 Ghana	1,330	3,720	41	53	19	7	41	40
23 Senegal	810	2,570	25	21	18	26	56	54
24 Chad	240	320	47	—	12	—	41	—
25 Mozambique	—	—	—	—	—	—	—	—
Middle-income oil importers	—	—	24w	20w	35w	32w	41w	48w
26 Mauritania	160	700	32	34	36	21	32	45
27 Liberia	270	980	27	36	40	26	34	38
28 Zambia	1,040	3,350	14	14	54	38	32	48

29 Lesotho	50	300	65	23	5	22	30	55
30 Cote d' Ivoire	960	7,090	36	27	17	24	47	50
31 Zimbabwe	960	4,730	18	11	34	32	48	57
32 Swaziland	70	—	35	—	33	—	32	—
33 Botswana	50	890	34	10	19	43	47	48
34 Mauritius	190	910	16	14	23	25	61	61
Middle-income oil exporters	—	—	48w	26w	17w	35w	35w	39w
35 Nigeria	4,190	64,570	53	26	19	34	29	40
36 Cameroon	750	7,220	32	24	17	32	50	45
37 Congo, People's Rep.	210	2,110	19	7	19	55	62	38
38 Gabon	220	3,120	26	7	34	63	40	30
39 Angola	—	—	—	—	—	—	—	—
Sub-Saharan Africa	—	—	41w	29w	20w	29w	39w	42w

c. Figures are for 1966.
d. Figures are for 1982.

TABLE III : Annual Growth of Production

(in Percentage)

	GDP		Agriculture		Industry	
	1965-73	1973-83	1965-73	1973-83	1965-73	1973-83
Low-income economies	3.7w	2.1w	2.6w	1.5w	5 7w	1.6w
1 Ethiopia	4.1	2.3	2.1	1.2	6.1	2.6
2 Mali	3.1	4.1	2.9	5.0	5.2	0.6
3 Zaire	3.9	—1.0	—	1.4	—	—2.0
4 Burkina	2.4	3.5	—	1.3	—	5.1
5 Guinea-Bissau	—	2.1	—	—3.1	—	1.3
6 Niger	—0.8	5.2	—2.9	1.6	13.2	10.9
7 Malawi	5.7	4.2	—	4.1	—	4.2
8 Tanzania[b]	5.0	3.6	3.1	2.6	6.9	0.2
9 Burundi	4.8	3.6	4.7	2.3	10.4	8.3
10 Uganda[b]	3.6	—2.1	3.6	—1.6	3.0	—10.1
11 Togo	5.3	2.3	2.6	1.1	6.2	2.6
12 Gambia	4.5	3.2	4.5	—2.4	4.4	7.2

(Contd.)

13 Somalia[b]	—	2.8	—	3.5	—	1.1
14 Benin	2.2	4.8	—	2.7	—	6.9
15 Central African Rep.	2.7	1.0	2.1	2.4	7.1	1.8
16 Madagascar[b]	3.5	0.3	—	—0.2	—	—1.8
17 Rwanda	6.3	5.6	—	—	—	—
18 Guinea	3.0	3.1	—	2.4	—	6.7
19 Kenya	7.9	4.6	6.2	3.4	12.4	5.3
20 Sierra Leone	3.7	1.9	4.5	2.2	1.9	—2.9
21 Sudan	0.2	6.3	0.3	3.5	1.0	6.7
22 Ghana	3.4	—1.3	4.5	(.)	4.3	—7.0
23 Senegal	1.5	2.6	0.2	0.3	3.5	6.1
24 Chad	0.5	—5.8	—	—	—	—
25 Mozambique	—	—	—	—	—	—
Middle-income oil importers	5.4w	3.0w	3.2w	2.4w	5.6w	3.0w
26 Mauritania	2.6	2.5	—2.1	2.6	3.5	(.)
27 Liberia	5.5	0.2	6.5	2.0	6.2	—1.5
28 Zambia	3.0	0.2	—	1.4	—	—0.3
29 Lesotho	3.9	5.5	—	—	—	—
30 Cote d' Ivoire	7.1	4.7	3.7	4.0	8.8	7.3
31 Zimbabwe	7.3	1.8	—	1.2	—	(.)

32 Swaziland[b]	7.6	3.8	8.0	4.7	3.1	4.1
33 Botswana	14.8	10.2	6.4	—4.2	30.2	15.3
34 Mauritius	2.3	3.8	—	—3.4	—	4.8
Middle-income exporters	9.1w	1.6w	3.0w	—1.5w	18.9w	1.0w
35 Nigeria	9.7	1.2	2.8	−1.9	19.7	0.3
36 Cameroon	4.2	6.8	4.7	1.8	4.7	13.7
37 Congo, People's Rep.	6.8	7.9	4.1	0.4	9,3	12.7
38 Gabon	7.4	—4.9	—	—	—	—
39 Angola	—	—	—	—	—	—
Sub-Saharan Africa	6.6w	2.1w	2.9w	0.7w	14.6w	1.4w

b. Figures are for 1973-82.

TABLE IV : Level and Growth of Merchandise Trade

	Merchandise (millions of dollars)		Annual growth rate (percent)			
			Exports		Imports	
	Exports 1983	Imports 1983	1965-73	1973-83	1965-73	1973-83
Low-Income economies	8,512t	12.486t	2.2w	—3.8w	2.5w	—2.1w
1 Ethiopia	422	875	3.0	1.4	—0.2	2.7
2 Mali	106	344	13.1	5.1	8.5	3.9
3 Zaire	1,459a	953a	6.5	—8.7	9.6	—13.7
4 Burkina	99	288	—1.0	1.7	7.2	4.2
5 Guinea-Bissau	..	..	..	..	..	..
6 Niger	301a	443a	6.1	19.0	4.4	11.5
7 Malawi	220	312	3.8	2.8	6.4	—0.6
8 Tanzania	480a	1,134	0.9	—4.6	7.1	−2.7
9 Burundi	76	194	..	..	..	..
10 Uganda	354	340	0.2	—8.0	—2.5	1.9
11 Togo	242	284	4.4	3.5	6.6	7.4
12 Gambia	..	..	—3.2	—6.9	2.2	5.0
13 Somalia	163	422	6.7	7.3	1.4	0.0

(*Contd.*)

14 Benin	85	523	12.4	—1.4	13.2	4.5
15 Central African Rep.	106a	132a	—0.4	3.8	—0.5	2.5
16 Madagascar	329a	439a	5.4	—4.3	1.5	—2.5
17 Rwanda	80	279	6.3	2.6	4.6	12.9
18 Guinea	390	279	..	..	..	..
19 Kenya	975	1,274	3.8	—4.8	5.9	—4.6
20 Sierra Leone	202	174	2.2	—5.3	0 9	—5.0
21 Sudan	624	1,354	3.8	—1.5	4.9	1.3
22 Ghana	895	719	3.5	—6.4	−3.3	—8.0
23 Senegal	585	984	−1.3	−0.9	5.4	—1.2
24 Chad	58a	109a	—3.5	—3.1	18.7	−8.6
25 Mozambique	260	635	−7.9	−8.3	−8.9	—4.2
Middle-Income oil importers	5,298t	4,578t	4.4w	—0.8w	6.4w	−1.9w
26 Mauritania	246	227	9.7	0.5	15.4	−0.8
27 Liberia	841	415	8.9	—2.3	3.6	—4.3
28 Zambia	870	690	—0.3	—0.8	3.0	—7.3
29 Lesotho	..	..	..	..	..	..
30 Cote d' Ivoire	2.068	1,814	7.1	—1.4	7,8	0.1

31 Zimbabwe	1,273a	1,432a	..	..	..	..
32 Swaziland	..	..	..	..	..	..
33 Botswana	..	..	..	..	..	..
34 Mauritius	..	..	4 2	4.2	4.5	—0.4
Middle-Income oil exporters	23,529t	21,198t	7.4w	—6.1w	7.8w	10.9
35 Nigeria	17.509a	17,600	8.9	—6.2	8.9	13.6
36 Cameroon	1,708	1,226	4.2	3.9	6.3	5.1
37 Congo, Peoples's Rep.	887	806	—2.2	4.4	—2.3	12.0
39 Gabon	1,566a	799a	6.8	—3.1	7.9	3.8
39 Angola	1,859	768	5.4	13.3	8.3	3.3
Sub-Saharan Africa	37,338t	38,261t	5.8w	—5.0w	3.9w	3.3w

(a) Figures are for 1972.

TABLE V : External Public Debt and Debt Service

(Millions of Dollars)

	External public and publicly guaranteed debt outstanding and disbursed						Debt service	
	Official sources		Private sources		Total			
	1970	1984	1970	1984	1970	1984	1970	1984
Low-income economies	2,294.2t	25,070.8t	897.2t	5,295.6t	3,191.4t	30,366.4t	229.7t	2,044.6t
1 Ethiopia	140.2	1,202.7	28.7	181.5	168.9	1,384.2	21.1	84.1
2 Mali	231.5	930.0	6.1	30.0	237.6	960.0	0.7	17.4
3 Zaire	95.3	3,371.6	215.8	712.1	311.1	4,083.7	36.8	352.3
4 Burkina	20.5	369.6	0.3	37.9	20.8	407.5	1.9	21.8
5 Guinea-Bissau	—	126.5	—	22.9	—	149.4	—	3.3
6 Niger	31.2	521.9	0.5	156.0	31.7	677.9	2.3	66.6
7 Malawi	89.1	603.1	33.3	127.5	122.4	730.6	5.9	82.3
8 Tanzania	148.9	2,132.1	100.7	461.5	249.6	2,593.6	15,7	71.3
9 Burundi	5,8	310.5	1.5	23.9	7.3	334.4	0.6	17.3
10 Uganda	108.6	605.6	29.7	69.5	138.3	675.1	7.9	86.3
11 Togo	31.9	561.6	7.8	97.6	39.7	659.2	2.3	66.6
12 Gambia	5.1	128.0	(.)	32.9	5.1	160.9	0.1	7.3

13 Somalia	74.9	1,173.8	2.2	59.2	77.1	1,233.0	0.9	26.8
14 Benin	29.2	300.6	11.3	280.9	40.5	581.5	1.7	38.3
15 Central African Rep.	17.9	197.6	6.1	26.8	24.0	224.4	2.9	12.1
16 Madagascar	84.9	1,239.2	7.6	397.2	92.5	1,636.4	6.9	116.6
17 Rwanda	1.5	243.9	0.4	0.0	1.9	243.9	0.3	6.0
18 Guinea	275.2	989.7	36.7	178.5	311.9	1,168.2	14.5	105.1
19 Kenya	234.2	2,148.9	84.3	484.5	318.5	2,633.4	27.4	348.4
20 Sierra Leone	32.4	252.6	27.0	89.0	59.4	341.6	12.0	16.2
21 Sudan	260.2	4713.1	46.2	945.8	306.4	5,658.9	34.6	107.3
22 Ghana	270.6	996.1	223.9	126.4	494.5	1,122.5	23.8	81.0
23 Senegal	80.5	1,357.9	19.6	197.2	100.1	1,555.1	6.7	92.7
24 Chad	24.6	85.1	7.5	23.9	32.1	109.0	2.7	2.5
25 Mozambique	—	509.1	—	532.9	—	1,042.0	—	215.0
Middle-income oil importers	559.2t	7,040.9t	830.0t	4,888.0t	1,389.2t	11,928.9t	135.0t	1,264.3t
26 Mauritania	19.4	1,047.9	7.9	122.7	27.3	1,170.6	3.3	42.2
27 Liberia	124.0	595.9	34.9	160.8	158.9	756.7	17.6	42.2
28 Zambia	119.4	2,149.4	503.1	629.3	622.5	2,778.7	59.0	113.1

(Contd.)

29 Lesotho	7.6	121.8	0.5	12.5	8.1	134.3	0.5	21.1
30 Cote d'Ivoire	144.2	1,986.8	112.0	2,847.8	256.2	4,834.6	38.4	641.0
31 Zimbabwe	88.1	482.3	144.6	963.4	232.7	1,445.7	9.4	276.4
32 Swaziland	20.9	169.6	16.1	8.4	37.0	178.0	3.3	20.2
33 Botswana	14.2	239.4	0.6	36.7	14.8	276.1	0.6	33 1
34 Mauritius	21.4	247.8	10.3	106.7	31.7	354.2	2.9	75.0
Middle-income oil exporters	699.7t	4,858.6t	145.9t	11,674.8t	845.6t	16,533.4t	84.4t	4,116.9t
35 Nigeria	385.1	2,182.7	94.4	9,632.7	479.5	11,815.4	55.7	3,162.8
36 Cameroon	119.6	1,365.1	11.6	372.8	131.2	1,737.9	8.6	221.5
37 Congo, People's Rep.	128.0	824.1	16.0	571.5	144.0	1,395.6	8.8	250.9
38 Gabon	67.0	257.0	23.9	467.5	90.9	724.5	11.3	252.8
39 Angola	—	229.7	—	630.3	—	860.0	—	228.9
Sub-Saharan Africa	3,553.1t	36 970.3t	1,873.1t	21,858.4t	5,426.2t	58,828.7t	449.1t	7,425.8t

TABLE VI : **Population Growth and Projections**

	Annual growth of population (percent)			Population (millions)		
	1965-73	1973-83	1980-2000	1983	1990	2000
Low-income economies	2.6w	2.8w	3.1w	253t	312t	423t
1 Ethiopia	2.6	2.7	2.6	41	48	64
2 Mali	2.6	2.5	2.5	7	9	11
3 Zaire	2.1	2.5	3.1	30	37	50
4 Burkina	2.0	1.9	2.0	6	7	9
5 Guinea-Bissau	1.2	4.3	2.1	1	1	1
6 Niger	2.6	3.0	3.2	6	8	11
7 Malawi	2.8	3.0	3.1	7	8	11
8 Tanzania	3.1	3.3	3.4	21	27	37
9 Burundi	1.4	2.2	2.9	4	5	7
10 Uganda	3.4	2.8	3.3	14	18	25
11 Togo	2.8	2.6	3.2	3	4	5
12 Gambia	3.0	3.6	2.7	1	1	1

13 Somalia	3.5	2.8	3.0	5	6	8
14 Benin	2.6	2.8	3.1	4	5	6
15 Central African Rep.	1.6	2.3	2.7	2	3	4
16 Madagascar	2.4	2.6	3.1	9	12	16
17 Rwanda	3.1	3.4	3.4	6	7	10
18 Guinea	1.8	2.0	2.1	6	7	8
19 Kenya	3.7	4.0	3.9	19	25	36
20 Sierra Leone	1.7	2.1	2.3	4	4	5
21 Sudan	2.6	3.2	2.8	21	25	33
22 Ghana	2.2	3.1	3.5	13	17	23
23 Senegal	2.4	2.8	2.9	6	8	10
24 Chad	1.8	2.1	2.4	5	6	7
25 Mozambique	2.3	2.6	2.9	13	16	22
Middle-income oil importers	3.3w	3.5w	3.4w	31t	40t	55t
26 Mauritania	2.3	2.2	2.6	2	2	3
27 Liberia	2.8	3.3	3.1	2	3	3
28 Zambia	3.0	3.2	3.3	6	8	11
29 Lesotho	2.1	2.5	2.6	1	2	2
30 Cote d' Ivoire	4.6	4.6	3.6	9	13	17
31 Zimbabwe	3.4	3.2	3.6	8	10	14

(Contd.)

32 Swaziland	3.1	3.4	3.3	1	1	1
33 Botswana	3.0	4.5	3.5	1	1	2
34 Mauritius	2.0	1.4	1.6	1	1	1
Middle-income oil exporters	2.5w	2.7w	3.3w	114t	143t	197t
35 Nigeria	2.5	2.7	3.3	94	118	163
36 Cameroon	2.4	3.1	3.2	10	12	17
37 Congo, People's Rep.	2.6	3.1	3.7	2	2	3
38 Gabon	0.2	1.4	2.6	1	1	1
39 Angola	2.2	2.6	2.8	8	10	13
Sub-Saharan Africa	2.6w	2.9w	3.2w	398t	496t	675t

Selected Bibliography

A. Hoben, *Land Tenure Among the Amhara of Ethiopia*, University Press, Chicago, 1975.

Abebe Teferri, *Unorganised Money Market in Ethiopia*, Haile Salassie I University, Faculty of Arts, Addis Ababa, 1966.

Abera Moltot, *The Need for Public Participation in the Execution of Land Reform Laws*, Seminar Proceedings in Agrarian Reform, A Policy Oriented Study, Vol. II, Addis Ababa, 1970.

Addis Ababa University, *Agricultural Mechanisation Suited to Ethiopian Conditions* Department of Economics, May, 1980.

Agency for International Development, *Findings and Implications, Spring, Review of Land Reform*, Department of State, Washington, 1970.

Albert Hirschman, *The Strategy for Economic Development*, New Haven, Yale University Press, 1964.

Alemayehu Lirenso, "*State Policies in Production, Marketing and Pricing of Food Grains*," The Case of Ethiopia, *African Development*, Vol. VII, No. 1 1983.

Alula Abate and Tesfaye Teklu, *Land Reform and Peasant Association in Ethiopia, A Case Study of Two Widely*

Differing Regions, World Employment Programme Research Working Paper, ILO, Geneva, 1979.

Alula Abate, *Peasant Association and Collective Agriculture in Ethiopia*, Promise and Performance, 1983.

Arthur W. Lewis, *The Theory of Economic Growth*, George Allen and Unwin Ltd., London, 1961.

Aster Akalu, *The Process of Land Nationalisation in Ethiopia*, Department of Sociology, Lund University, Lund, 1982.

A.K.S. Lambton, "Ethiopia : An Approach to Land Reform", Bulletin School of Oriental African Studies, London, 34 (2), 1971.

Assefa Bequele and Eshetu Chole, *A Profile of the Ethiopian Economy*, Addis Ababa, London, 1969.

Bakle Endeshaw, *Theoretical Analysis of the Effects of Tenure on Agricultural Productivity in the Empire of Ethiopia*, B.S. Imperial College of Agriculture and Mechanical Arts, Alammaya, 1957.

Basic Documents of the Ethiopian Revolution, Addis Ababa, 1977.

David A. Talbot, *Contemporary Ethiopia*, Philosophical Library, New York, 1952.

David A. Talbot, *Haile Sellassie I Silver Jubilee*, W.P. Van Stockum & Zoon, The Hauge, 1955.

Declaration of Economic Policy of Socialist Ethiopia, Addis Ababa, 1975.

Department of Land Tenure, Ministry of Land Reform and Administration, Annual Report, 1968-69.

Dessalegen Rahmato, *Agrarian Reform in Ethiopia*, Scandinavian Institute of African Studies, Uppsala, 1984.

Detsa Bishaw, *Agricultural Development in Ethiopia*, Haile Selassie I University, Addis Ababa, 1968.

Documents and Resolutions of the 2nd COPWE Congress, Addis Ababa, January 1983.

E. Ullendorf, *The Ethiopians, An Introduction to Country and People*, Oxford University Press, 1960.

E.W. Luther, *Ethiopia To-day*, Stanford, California, 1958.

E. Westphal, *Agricultural System in Ethiopia*, Agricultural Research Report No. 826, 1975.

ECA/FAO, *Organisation and Administration of Agricultural Extension Services in Ethiopia*, Agricultural Division, Addis Ababa, May 1969.

FAO, Commodity Bulletin No. 33, *The World Coffee Economy*, Rome, 1961.

FAO/UN, *Information on Land Reforms—Land Settlement and Co-operatives*, Rome, 1963.

FAO/UN, *Land Reform in Italy—Achievements and Prospectives*, Rome, 1961.

FAO/UN, *Report to the Imperial Ethiopian Government on Economic Policies for Expanding Agricultural Production*, Expanded Technical Assistance Programme, Report No. 926, Rome, 1958.

FAO/UN, *Report to the Government of Ethiopia—Land Tenure and Landlord-Tenant Relationships*, United Nations Development Programme, FAO No. TA 2651, Rome, Based on the work of Dr. H.S. Man.

FAO/UN, *State of Food and Agriculture*, Rome, 1961.

FAO/IBRD, *Co-operative Programme in Ethiopia*, A Draft Scheme for the Development of Agricutural Credit and Marketing in the Woreda of Sodo and Bolosso, December 1966.

Fassil G. Kiros, *Education for Integrated Rural Development in Ethiopia*, Institute of Development Research, Report No. 19, Addis Ababa, 1983.

Fassil G. Kiros, *An Investigation into the Problem of Fragmentation of Agricultnral Land in Ethiopia in the Post-1975*, Research Report No. 31, IDR, Addis Ababa University, 1980.

Fred Halliday and Maxine Molyneux, *The Ethiopian Revolution*, British Library, London, 1981.

G.A. Lipsky, *Ethiopia : Its People, Its Society, Its Culture*, Hraf Press, New Haven, 1962.

Georgi Galperin, *Ethiopia : Population, Resources, Economy*, Progress Publishers, Moscow, 1981.

Girma Zawdie, "What Marketing Means to the Agriculturists," *The Ethiopian Herald*, Addis Ababa, September, 1969.

Haile Atsebcha, *Economics and Statistics Department*, Ministry of Agriculture, *Comments on the work of Dr. Willis Eichberger entitled Food Situation in Ethiopia*.

Haile S. Belay, *Supporting Institutions for Effective Land Reforms with Special Reference to Local Government*, Addis Ababa, 1969.

Haile W. Michael,*Zemecha, An Attempt to Rural Transformation in Ethiopia*, (Draft) 1975.

Henis Angerer, *Report and Proposal Concerning the Possibilities of Agricultural Irrigation in Ethiopia*, Addis Ababa, 1966.

Hari Saran Chhabra, "Socialist Ethiopia", The Hindustan Times January 7, 1986.

H.P. Huffnagel, *Agriculture in Ethoipia*, FAO/UN, Rome, 1961.

IAR, *Outstations in Holeta Genet*, Research Station, Progress Report, April 1975 to March 1976, Addis Ababa.

IBRD/IDA, *The Economy of Ethiopia, Vol. II—Agriculture*, Africa Development, 1967, (Restricted Report).

Imperial Ethiopian Government, *Five Year Development Plan, 1957-61*, Addis Ababa.

Imperial Ethiopian Government, *Second Five Year Development Plan, 1963-67*, Addis Ababa, 1962.

Imperial Ethiopian Government, *Third Five Year Development Plan, 1968-73*, Addis Ababa, 1968.

Imperial Ethiopian Government, *A Proclamation to Provide for the Additional Agricultural Income Tax*, Addis Ababa, 1967 ; *A Decree to Provide for the Educational Income Tax*, 1959 ; *A Decree to Provide for the Creation of Farm Workers' Co-operatives*, Addis Ababa, 1960 ; *A Proclamation to Provide for the Formation of the Co-operative Sector*, Addis Ababa, 1966.

Information and Propaganda Committee of the Fifth Anniversary of the Ethiopian Revolution, Addis Ababa, 1979.

Jack Sheperd, *The Politics of Starvation*, New York, 1975.

J. Lawrence and H.S. Man, "FAO Land Policy Project", *Ethiopian Observer*, Vol. 9, No. 4, 1966.

J. Markakis, *Ethiopia, Anatomy of a Traditional Polity*, Oxford, Clarendon Press, 1974.

John Holmberg, *Grain Marketing and Land Reform in Ethiopia*, Report No. 41, Scandinavian Institute of African Studies, Uppsala, 1977.

Changes in the Legal Orders," African Studies Review, Mass, Vol. XXII, No. 1, April, 1979.

Peter Schwab, *Ethiopia, Politics, Economics, Society*, Frances Printer, London, 1985.

PMAC , *A Proclamation to Provide for the Public Ownership of Rural Lands*, Proclamation No. 31 of 1975, Addis Ababa, April 1975.

——*A Proclamation to Empower Peasant Associations*, Proclamation No. 71 of 1975, Addis Ababa, December 1975.

——*A Proclamation on Ownership of Urban and Extra Houses*, Proclamation No. 47 of 1975, Addis Ababa, July 1975

——*A Proclamation to Provide for the Rural Land Use Payment and Agricultural Activities Income Tax*, Proclama-on No. 77 of 1976, Addis Ababa, January, 1976.

——*A Proclamation on the Establishment of the AMC*, Proclamation No. 105 of 1976 and also issued on 20th November 1979, Addis, Ababa,

——*A Proclamation on the Formation of the Revoloutionary Development Committe*, Proclamation No. 115 of 1977, Addis Ababa, September 1977

——*A Proclamation on the Formation of Nation-Wide Peasant Associations*, Proclamation No. 130 of 1977, Addis Ababa, September 1977.

——*A Proclamation on Restructuring Peasant Association*, Proclamation No. 223 of 1982, Addis Ababa, May 1982.

R. Lamb and D. Milas, "Soil Erosion, Real Cause of Ethiopian Famine, Environmental Conservation, Vol. 10, No. 2, 1983.

R. Greenfield, *Ethiopia—A* New Political History, Pall Hall Press, London, 1965.

Ragnar Nurkse, *Problems of Capital Formation in Under—developed Countries*, Oxford University Press, Oxford, 1958.

Raymound P. Christersen, ''Agricultural Progress in Less Developed Countries in Economic Development" Published in "*The Modernisation of Farming*" Iowa State University Press, Iowa, USA. 1965.

——*A Guide to Development Planning*, D.No. PL-G-8.

——*Proposed Land Reform Development Programme and Its Estimated Effects on Productivity of Agriculture*, D.No.AG-LR-16, 1959.

——*Report on Possibilities of Introducing Co-operative Associations in the Agricultural Development in Ethiopia*, D.No. AG-LR-32.

——*Regional Aspects of National Planning in Ethiopia*, Part, I and II, Addis Ababa, 1968.

——*Third Five Year Development Plan—1961-65 EC*, (*Second Draft*), *Agricultiure, Forestry, Fishery and Water Resources*, Addis Ababa, November, 1968.

M.M. Bobbler, *Karl Marx's Interpretation of History*, Cambridge University Press, 1950

M. Perham, *The Government of Ethiopia*, Faber and Faber, London, 1969

M. Stahl, Ethiopia, *Political Contraditions in Agricultural Development* Stockholm, 1974.

M. Stahl, *New Seeds in Old Soil, A Study of the Land Reform Process in Western Wollega, Ethiopia, 1975-76*. Scandinavian Institute of African Studies, Uppsala, 1977.

Mulugetta Beyene, "Strengthening of Co-operatives Movement", *The Ethiopian Herald*, January, 29 1982.

M. Tucoric, *Thesis on the Development and Organisation of Agricultural Production in Ethiopia*, Addis Ababa.

National Coffee Board, *The Story of the National Coffee Board of Ethiopia 1957-68*, Addis Ababa, July, 1969.

National Disaster Aid Coordinating Central Committee, *Settlement Lasting Solution to People in Drought Affected Area of Ethiopia* Addis Ababa, 1979.

Office of the Population and Housing Commission, *Population and Housing Census*, Preliminary Report, Vol. I, No. I, 1984.

P. Gilkes, The Dying Lion, *Feudalism and Modernisation in Ethiopia*, London, 1975.

Paul Brietzke, "Land Reform in Revolutionary Ethiopia," *The Journal of Modern African Studies*, Vol. 14, No. 4, December, 1976,

Peter Koehn, "Ethiopia, Famine, Food Production and

Changes in the Legal Orders," African Studies Review, Mass, Vol. XXII, No. 1, April, 1979.

Peter Schwab, *Ethiopia, Politics, Economics, Society*, Frances Printer, London, 1985.

PMAC , *A Proclamation to Provide for the Public Ownership of Rural Lands*, Proclamation No. 31 of 1975, Addis Ababa, April 1975.

——*A Proclamation to Empower Peasant Associations*, Proclamation No. 71 of 1975, Addis Ababa, December 1975.

——*A Proclamation on Ownership of Urban and Extra Houses*, Proclamation No. 47 of 1975, Addis Ababa, July 1975

——*A Proclamation to Provide for the Rural Land Use Payment and Agricultural Activities Income Tax*, Proclama-on No. 77 of 1976, Addis Ababa, January, 1976.

——*A Proclamation on the Establishment of the AMC*, Proclamation No. 105 of 1976 and also issued on 20th November 1979, Addis, Ababa,

——*A Proclamation on the Formation of the Revoloutionary Development Committe*, Proclamation No. 115 of 1977, Addis Ababa, September 1977

——*A Proclamation on the Formation of Nation-Wide Peasant Associations*, Proclamation No. 130 of 1977, Addis Ababa, September 1977.

——*A Proclamation on Restructuring Peasant Association*, Proclamation No. 223 of 1982, Addis Ababa, May 1982.

R. Lamb and D. Milas, "Soil Erosion, Real Cause of Ethiopian Famine, Environmental Conservation, Vol. 10, No. 2, 1983.

R. Greenfield, *Ethiopia—A* New Political History, Pall Hall Press, London, 1965.

Ragnar Nurkse, *Problems of Capital Formation in Under—developed Countries*, Oxford University Press, Oxford, 1958.

Raymound P. Christersen, "Agricultural Progress in Less Developed Countries in Economic Development" Published in "*The Modernisation of Farming*" Iowa State University Press, Iowa, USA. 1965.

– –*A Guide to Development Planning*, D.No. PL-G-8.

–––*Proposed Land Reform Development Programme and Its Estimated Effects on Productivity of Agriculture*, D.No.AG-LR-16, 1959.

–––*Report on Possibilities of Introducing Co-operative Associations in the Agricultural Development in Ethiopia*, D.No. AG-LR-32.

–––*Regional Aspects of National Planning in Ethiopia*, Part, I and II, Addis Ababa, 1968.

–––*Third Five Year Development Plan—196I-65 EC*, (*Second Draft*), *Agricultiure, Forestry, Fishery and Water Resources*, Addis Ababa, November, 1968.

M.M. Bobbler, *Karl Marx's Interpretation of History*, Cambridge University Press, 1950

M. Perham, *The Government of Ethiopia*, Faber and Faber, London, 1969

M. Stahl, Ethiopia, *Political Contraditions in Agricultural Development* Stockholm, 1974.

M. Stahl, *New Seeds in Old Soil, A Study of the Land Reform Process in Western Wollega, Ethiopia, 1975-76*. Scandinavian Institute of African Studies, Uppsala, 1977.

Mulugetta Beyene, "Strengthening of Co-operatives Movement", *The Ethiopian Herald*, January, 29 1982.

M. Tucoric, *Thesis on the Development and Organisation of Agricultural Production in Ethiopia*, Addis Ababa.

National Coffee Board, *The Story of the National Coffee Board of Ethiopia 1957-68*, Addis Ababa, July, 1969.

National Disaster Aid Coordinating Central Committee, *Settlement Lasting Solution to People in Drought Affected Area of Ethiopia* Addis Ababa, 1979.

Office of the Population and Housing Commission, *Population and Housing Census*, Preliminary Report, Vol. I, No. I, 1984.

P. Gilkes, The Dying Lion, *Feudalism and Modernisation in Ethiopia*, London, 1975.

Paul Brietzke, "Land Reform in Revolutionary Ethiopia," *The Journal of Modern African Studies*, Vol. 14, No. 4, December, 1976,

Peter Koehn, "Ethiopia, Famine, Food Production and

——*Co-operative Societies Manual*, Community Development Department, Addis Ababa. 1968.

——*Community Development in Ethiopia, A Brief Resume of the Proceedings of the Third Seminar of District Development Officers*, Addis Ababa, 1965.

——*Labour, Co-operatives, School Welfare and Community Development in Ethiopia. A Brief Review*, 1957-64, November. 1964.

——*Production Plan for Arba-Minch Farm* 1969-70, Department of Agro-Industrial Development, Addis Ababa, June, 1169.

——*Proposals for Third Five Year Plan, 1968-73*, Addis Ababa, 1968.

Ministry of Planning and Development/Planning Commission/ Office of the Planning Board, *Agriculture : Second Five Year Plan*, Documentation No. AG-G-17.

——*Agriculture : Annual Plans*, D. No. AG-AN-4.

——*Agriculture, Wildlife, Forestry and Fishery*, D. No. AG-G-16.

——*Co-operative Possibilities in Ethiopia*, D. No. AG-LR-35.

——*Development of Agriculture Fishery, Forestry During the Second Five Year Development Plan*, D. No. AG-G-1 May, 1962.

——*Documentation of the Five Year—Appendix Section—Agriculture and Forestry*, D. No. AG-G-8.

——*Effects of Extension Service on Productivity of Agriculture*, D. No. AG-G-29.

——*Evaluation Report on the Economic and Social Development During the First Five Year Plan*, D. No.Eco. 6.

——*Implementation Report of 1956* EC, D. No. AG-AN-6, November, 1964.

——*Implementation of the Five Year Plan, Evaluation for the period 1957-59*, November. 1959.

——*Instructions for the Elaboration of Argicultural Programme*, D.No. Ag-G-8.

——*Methodology for the Estimation of Agricultural Output in* 1955 EC, November, 1962.

——*Land Tenure Problems of Ethiopia*, D.No. AG-LR-3.

——*Land Settlement Programme*, D.No. AG-LR-19, 1962.

John W. Mellor, *The Economics of Agricultural Development*, Cornell University Press, New York, 1966.

J.W. Bruce, *Ethiopia, Nationalisation of Rural Land Proclamation*, University of Wisconsin Land Tenure Centre, Madison, 1971.

J.M. Cohen, *Revolution and Land Reform in Ethiopia*, Rural Development Occasional Paper No. 6, Cornell University, New York, 1976.

K.A. Cherian, *Ethiopia To-day, An Up-to-date Illustrated Review of Economic Development*, Central Printing Press, Addis Ababa, June, 1969.

Kool Rudolf, *Tropical Africa and Economic Development*, Wagenigen, 1960.

L. Bondestam, "*Notes on Foreign Investment in Ethiopia*" *in Multinationl Firms in Africa*, Uppsala, 1975.

Land Policy Project, *Tenancy Reforms in Ethiopia*, Addis Ababa, 1964.

Mengistu Woube, *Problems of Land Reform Implementation in Rural Ethiopia—A Case Study of Dejen and Wolmera Districts*, Geografiska Regionstudier Nr. 16, Uppsala, 1986.

Ministry of Agriculture, *2nd Five Year Development Plan, 1962-67*, Addis Ababa.

——*Cotton in Ethiopia*, Addis Ababa, 1956.

——*Review of Agrarian Reform and Rural Develapment in Ethiopia*, Prepared for high level mission on the follow-up to the World Conference on Agrarian Reform and Rural Development, Addis Ababa, April 1982.

——*Towards Generalisation of Integrated Rural Development in Ethiopia*, Addis Ababa, December 1979.

Ministary of Information, *Agriculture in Ethiopia, Book II*, Addis Ababa, 1964.

Ministry of Land Reform and Administration, Department of Land Tenure, *Annual Report, 1968-69*, Addis Ababa.

Ministry of National Community Development and Social Affairs, *A Guide for Village Level Workers*, Addis Ababa October, 1967.

——*Arba-Minch Pilot Farm, Standard Conditions of Settlement* Scheme, Addis Ababa.

Richard Pankhurst, *An Introduction to the Economic History of Ethiopia from early times to 1800*, Stapler Printer Ltd., 1961.

——*Economic History of Ethiopia, 1800-1932*, Haile Selassie I University, Addia Ababs, 1968.

——*State and Law in Ethiopian History*, Haile Selassie I University, Addis Ababa, 1966.

SIDA—Project Preparation Team, *Rigional Development Project in Ethiopia*, Report No. I, Part I and II, Appendix (Summary).

S. Krikor, *Memorandum Concerning the Land Reforms in Ethiopia*, Addis Ababa, June 1961

Simon Kuznets, *Lecture on Economic Growth*, John Hopkins, Illiniois, 1954

Tadesse Araya, *Land Tenure and the Need for Land Reforms in Ethiopia*, Institute of Social Studies, The Hauge, The Netherland, September 1968,

T.W. Schultz, *Transforming Traditional Agriculture*, New Haven Yale University Press, 1966.

The Technical Agency, IEG, *An Agricultural Credit Program for Ethiopia*, Stanford Research Institute, California, January, 1969.

——*Ethiopia's Export Trade in Major Agricultural Commodities*, SRI, June 1969.

——*Marketing of Grains and Pulses in Ethiopia*, SRI, 1969.

——*Potential Fertiliser Demand in Ethiopia*, SRI, 1968.

——*Production of Grains and Pulses in Ethiopia*, SRI, 1969.

——*System Analysis Methods for Ethiopian Agriculture*, SRI, 1968.

——*The Development of Oilseeds Industry in Ethiopia*, SRI, 1964.

——*The Resources and Economy of Ethiopia*, SRI, 1969.

The World Bank, *Financing Adjustment with Growth in Sub-Saharan Africa, 1986-90*, Washington, 1986.

United Nations, *Community Development and Economic Development, Part I—A Study of the Contribution of Rural Community Development Programmes to National Economic Development in Asia and the Far East*, Bangkok, 1960.

United Nations, Economic and Social Council, *Progress in Land Reform*, Fourth Report, E/4020/Addi. I, 1955.

UN/FAO., *Report of the Exploratory Mission to Ethiopia in connection with a Possible Agricultural Economic Survey*, Rome, 1965.

United Nations *Conference on the Least Developed Countries*, Addis Ababa, 1981.

UN/FAO., *African Agricultural Development*, New York, 1966, E/CN. 14/342.

USA Operation Mission—*Point 4-The Agriculture of Ethiopia*, Vol. 1., January, 1954.

USAID/Ethiopia, *Agro-Industial Sector Study-Agriculture*, Addis Ababa, June 1968.

——*Shashamanne Farm Development Project*, Agricultural Division, Addis Ababa, 1969.

Willence Johnson, "Food and Politics—A Case Study of Ethiopia," Horn of Africa, Vol. II, No. 2, 1978.

William H. Nicholis, *The Place of Agriculture in Economic Development*, Edited by Carl Eicher and Lawrance Witt, MaGraw-Hill Book Company, New York, 1964.

W.H. Wcoultas, *A Report of the Investigation and Possibilities for Establishing Tea Industry in Ethiopia*, Addis Ababa, 1968.

W.W. Rostow, *Stages of Economic Growth*, Cambridge University Press, Cambridge, 1960.

Willis G. Eichberger, *Food Production and Utilisation in Ethiopia*, Addis Ababa, May, 1968.

——*Price Policy for Agricultural Commodities*, Addis Ababa, November, 1969.

——*Total Food Demand in Ethiopia in 1972*, Addis Ababa, 1969.

World Bank, *Accelerated Development in Sub-Saharan Africa—An Agenda for Action*, Washington, 1981.

Zoltan Gyenge, *Ethiopia on the Road of Non-Capitalist Development* Institute for World Economics of the Hungarian Academy of Science, Budapest, 1976.

JOURNALS AND NEWSPAPERS

Addis Ababa Chamber of Commerce, Trade Directory and Guide Book of Ethiopia, Addis Ababa,

African Research Bulletin, Economic, Financial and Technical Series, Exeter, Devon, England.

Central Statistical Office, Ethiopia—Statistical Absract, Addis Ababa,

Central Statistical Office, Reports on Different Surveys, Addis Ababa.

Commerical Bank of Ethiopia, Annual Reports. Addis Ababa.

Development Bank of Ethiopia, Annual Reports, Addis Ababa.

Ethiopian Chamber of Commerce, Ethiopian Trade Journal, Addis Ababa.

Ethiopian Observer, Published in Ethiopia and Britain.

Haile Selassie I University, Journal of Ethiopian Studies, Addis Ababa.

Ministry of Information and Guidance, Quarterly YEKATIT.

Ministry of Commerce and Industry, Ethiopian Economic Review, Addis Ababa.

National Bank of Ethiopia, Annual Reports, Addis Ababa.

The Economist Intelligence Unit, Quarterly Economic Review of Ethiopia, London.

The Ethiopian Herald, Addis Ababa.

Index